Taxation of
Mineral Resources

Books from
The Lincoln Institute of Land Policy

The Lincoln Institute of Land Policy is a school that offers intensive courses of instruction in the field of land economics and property taxation. The Institute provides a stimulating learning environment for students, policymakers, and administrators with challenging opportunities for research and publication. The goal of the Institute is to improve theory and practice in those fundamental areas of land policy that have significant impact on the lives and livelihood of all people.

Constitutions, Taxation, and Land Policy
Michael M. Bernard

Constitutions, Taxation, and Land Policy—Volume II
Michael M. Bernard

Federal Tax Aspects of Open-Space Preservation
Kingsbury Browne

Taxation of Nonrenewable Resources
Albert M. Church

Taxation of Mineral Resources
Robert I. Conrad and R. Bryce Hool

Incentive Zoning
Jerald S. Kayden

Building for Women
Edited by Suzanne Keller

State Land-Use Planning and Regulation
Thomas G. Pelham

The Art of Valuation
Edited by Arlo Woolery

Taxation of Mineral Resources

Robert F. Conrad
Duke University

R. Bryce Hool
State University of New York
at Stony Brook

LexingtonBooks
D.C. Heath and Company
Lexington, Massachusetts
Toronto

Library of Congress Cataloging in Publication Data

Conrad, Robert F
  Taxation of mineral resources.

  Bibliography: p.
  Includes index.
  1. Mines and mineral resources—Taxation—United States. 2. Mining industry and finance—Taxation—United States. I. Hool, R. Bryce, joint author. II. Title.
HJ4169.C66   336.2'7833385'0973   80-8392
ISBN 0-669-04104-1

Copyright © 1980 by D.C. Heath and Company

Published simultaneously in Canada

Printed in the United States of America

International Standard Book Number: 0-669-04104-1

Library of Congress Catalog Card Number: 80-8392

To Helen and Emily

Contents

List of Figure
and Tables

Introduction
and Summary

There has been a substantial increase in recent years in the level of taxation imposed on mining firms by state and local governments. This increase can be attributed to three factors. First, there has been a heightened awareness that resources are limited in quantity and, consequently, there is a need to conserve the resource base. Second, the environmental damage resulting from mining operations, notably the deterioration of water quality and land abuse, has brought demands for just compensation. Third, significant price increases for some minerals have often been viewed by states as an opportunity to collect additional tax revenue.

It is important to understand the implications of this taxation and, in particular, the effects associated with each of the various forms of taxes that are being applied. Accordingly, the broad aim of this book is to provide a comprehensive economic analysis of the effects of mining taxation on the extraction of mineral resources. Although we shall deal specifically with the taxes imposed by state and local governments in the United States, the conclusions will have more general applicability, since the essential forms of the taxation are applied universally.

On the basis of this analysis, we offer a set of recommendations for tax policy. The primary objective of this design is to minimize the distortionary incentives created by the taxation. From a practical standpoint, however, one must also recognize the degrees of difficulty in the administration of the various taxes.

The analysis and recommendations will reflect the premise that, for tax purposes, a mining operation should be treated in a neutral manner, that is, like any other form of economic activity. We regard this premise as appropriate for two reasons. First, in the absence of proved market failures, a nonneutral tax treatment of the mining sector creates an inefficient allocation of resources. Too many or too few resources devoted to mining will be costly to states in terms of jobs and output, either in mining or some other activity. Second, if the tax system is nonneutral, it may be difficult to detect whether market failure is due to the nature of the activity or to relative distortions induced by the tax system. If, for instance, it is determined that a state's resource base is being depleted too fast, an output tax may be justified. However, if the resources are being depleted too fast because of an excessive property tax, for example, an additional tax may succeed only in deterring further development. It is therefore appropriate for states to move toward a neutral tax policy and then to evaluate any alternatives from that position.

The analysis is also based on the assumption that mining firms respond, in the short run, to changes in prices and costs and, in the long run, to changes in the net-of-tax rate of return on investment. A change in tax policy will always have an impact on mineral development. An increase in the mining industry's tax burden will result in slower growth of mineral development in the state and, consequently, lower output and job creation. To expect otherwise is to exaggerate the realizable benefits of taxation.

The study is presented in four chapters. The first two chapters establish the context for the economic analysis. In chapter 1 we set out the major categories of mining taxation and discuss salient features of the several forms within each category. References to the methods adopted by particular states are supplemented by a detailed tabular survey of the taxes applied in twenty-two states with significant mining sectors.

In chapter 2 we describe the main elements of the mining process itself. Recognition of the structure of mining decisions is essential to the relevance of economic analysis and policy prescription. Of particular importance here is the interaction of economic and geological factors.

The core analysis of the impact of taxes on a mining firm's investment, development, and extraction behavior is presented in chapter 3. We focus initially on the individual effects of the taxes and then consider the net effects of different taxes applied in combination. To illustrate the major influences, we construct numerical examples that display the variations in a firm's response to different taxes, in otherwise identical economic and geological conditions, as well as the variations in response to a given tax as economic and geological conditions vary.

Chapter 4 begins with an illustrative computation of the net quantitative effect of a state's tax system and then discusses tax shifting. It concludes with a presentation of recommendations for tax policy.

We now summarize briefly the main conclusions from the analysis and our policy recommendations. The particular taxes considered can be grouped in three distinct categories: severance (output-related) taxes, property taxes, and profits (or income) taxes.

Severance taxes may be specified as a fixed nominal payment per ton of final output (that is, payment of a specified number of dollars per ton, whatever the price level); as a fixed nominal payment per ton of ore extracted, before processing; or as a proportion of sales revenue. Each variant is distinct in its effects. A per-unit tax on output creates a tendency to reallocate extraction from present to future and may also alter the time profile of the quality of ore selected for extraction. A per-unit tax on ore is also an inducement to defer extraction, but such a tax will not influence the quality profile. An ad valorem tax will not alter the quality profile but will induce a shift in extraction from present to future, or vice versa, according to whether prices are rising at a rate less than or greater than the rate of interest. All variants lead to an increase in the cutoff grade of ore, reducing the size of economically recoverable reserves—the phenomenon of tax-induced high grading.

All severance taxes, and indeed all taxes, reduce the rate of return on capital and thereby serve to discourage investment and development expenditures. This is the major long-run distortion of mining taxation in general, whose incidence across states is in relation to the total tax burden imposed by each state.

Property taxes, in practice, are also levied in various forms, few of which show any evident connection to the economic definition of property value. As administered, property taxes have the advantage of stable revenue. Their disadvantage is the practical difficulty of property assessment. A true property tax, based on the estimated value of reserves remaining in a deposit, will effectively subsidize and thereby accelerate extraction. At the same time, it will tend to lower the cutoff grade and so increase total extraction.

Profits taxes may be proportional (levied at a uniform rate) or progressive (levied at an increasing rate) and are typically contaminated by special deductions, such as depletion allowances. In the absence of deductions, a proportional profits tax is nondistortionary with respect to extraction, whereas a progressive tax will induce the mining firm to modify its profits profile in a manner that will depend on the time paths of prices and costs. This redistribution may be brought about by changes in the rates of extraction or in the qualities of ore extracted, or some combination of the two.

Depletion allowances serve to subsidize extraction. Cost depletion, a fixed nominal allowance per ton of ore extracted, acts as a negative per-unit severance tax. Extraction tends to be reallocated from future to present; cutoff grades will be lower and recovery higher than otherwise. Percentage depletion, a fixed proportion of current revenue, acts as a negative ad valorem severance tax. Accordingly, it encourages a shift in extraction from future to present, or vice versa, as prices are rising at less or more than the rate of interest. It also lowers the cutoff grade and increases total recovery.

The preceding array of potential distortions, together with the relative difficulties of administering the options, lead to the following prescription for the general design of mineral tax policy.

1. A proportional profits tax should be the cornerstone. Its primary advantages are recognition of cost as well as revenue; avoidance of a high-grading incentive; and ability, in conjunction with the individual income tax, to collect the resource rents. The income-tax package should include these provisions: (a) percentage depletion should not be allowed; (b) capitalization of exploration expenditures should be required; (c) development expenses should be subject to usual depreciation rules; (d) all other state and local taxes should be deductible; and (e) taxable income should include only income derived within the state.

2. A property tax is recommended for states that rely on their mining sector for a stable source of revenue. Administration of the proposed tax is likely to be proportionately less costly for such states. The property-tax base should be an estimate of present value, the future income stream being obtained by applying a historical average profit margin to estimates of reserves and production. The use of a moving historical average profit margin will reduce the incentive for high-grading and accelerated extraction.

3. Given the relative ease of administration, many states will probably impose a severance tax of some sort. If so, we recommend that it be an ad valorem royalty, since this does not bias the quality profile or induce any consistent quantity-profile bias. All severance taxes have the undesirable consequence of high-grading, but per-unit severance taxes also induce extraction-profile distortions. We recommend further that (a) the tax should be imposed on ore, prior to pro-

cessing; (b) ore prices should be quality-adjusted, with a standard price derived, if necessary, from the price of concentrate with a deduction for processing costs; and (c) the tax rate should be uniform, that is, independent of the price. The justifications for these proposals and for the rejection of the alternatives are detailed in the final chapter.

Acknowledgments

Financial support for the research for this book was generously provided by the Lincoln Institute of Land Policy. We wish to thank Arlo Woolery, director of the Lincoln Institute, for his encouragement of this research.

We are also grateful to all the state tax administrators for their helpful cooperation.

1   Mining Taxes

There is considerable variation across states in the types of mining taxation currently employed and even in the names they go by. To simplify the description and discussion of these taxes we divide them into three broad categories: output-related (or severance) taxes, profits (or income) taxes, and property taxes. This classification is also used in the subsequent analysis.

Output-Related Taxes

Output-related taxes are defined here as taxes imposed on a per unit basis on mineral output, either at the mouth of the mine or after concentration. They are most commonly known as severance taxes but are also referred to as production or mining privilege taxes. This type of tax takes two general forms: a fixed nominal amount per unit of output, and a fixed proportion of the value of output (an ad valorem tax).

    These taxes are popular means of raising revenue for several reasons. First, the administrative costs are relatively small when compared to other forms of taxation, on account of clearly defined units of taxation. No costs, depletion, or depreciation need be calculated and valuation problems are relatively minor. However, as Stinson (1977) notes, there are difficulties in assigning values for purposes of an output tax based on gross value. The problem arises because arm's-length market transactions may not occur. This is especially a problem in vertically integrated operations that purchase mineral output as an input into the production process (not necessarily in the same state).

    Second, the output tax is perceived as collecting part of the value of extracted resources. Unlike other economic operations, mineral extraction permanently reduces the natural wealth of the state. Thus output taxes are seen as providing a tangible link between the extrac-

1

tion process and the perceived resource rents. We question this argument in chapter 3.

Third, the revenue from such taxes, whose source is readily identifiable, is fairly easy to earmark for specific purposes. Like gasoline taxes used for highway trust funds, output taxes are sometimes placed in special funds for land reclamation and other projects identified with the mining process. The tax is thus viewed as a means of forcing the mines to "pay their own way" for public services and environmental damage. There is no inherent reason that other types of taxes cannot be similarly earmarked. However, the advantage of output taxes is that a legislature can link a tax to a specific purpose without allocating a fixed nominal amount of expenditure for the services in the budget.

Finally, it has been suggested that these taxes are passed through in the form of higher prices to consumers of finished products, and thus the tax is "exported" out of the state. This claim is examined in detail in chapter 4.

Output-related taxes are not without their difficulties. Chapter 3 shows that severance taxes favor large operations of high-grade deposits whose profits can easily absorb the tax. Second, they increase the cost of extraction regardless of the size of the operation and quality of the deposit, causing high-grading, that is, the bypassing of ores that could be profitably extracted in the absence of the tax. Third, they may alter the intertemporal extraction profile.

Table 1-1 compiles the types of output-related taxes in the states surveyed. It is evident that the taxes and rates vary widely across states for the same mineral. For instance, coal in Ohio is taxed at a rate of only 4¢ per ton, while coal in South Dakota is taxed at 50¢ per ton, plus a 1-percent increase for every 3-percent increase in the wholesale price index (WPI). The taxes also vary across minerals in the same state. This reflects in part the legislature's perception of cost differentials and the relative importance of a particular mineral in the state's economic base. (Other economic factors also play a role. For example, the recent decrease in the copper royalty in Arizona reflects a decline in copper prices.)

It is apparent that states are aware of the adverse incentives created by this output tax. Some states lower tax rates for small producers (Wyoming, Alaska, and Montana) as well as allowing rates to vary with quality differentials (Wyoming and Alaska). The need for

Table 1-1
Output-Related Mining Taxes

State/Mineral	Name	Base and Rate	Remarks
Alabama			
Oil and gas	Production	4% gross value	Paid in lieu of property taxes. Revenue distributed to state, counties, and cities by set formula.
Oil and gas	Conservation	2% gross value[a]	Revenue goes into state general fund.
Iron ore	Severance	3¢/long ton	Revenue goes into state general fund.
Coal	Severance	33.5¢/long ton	13.5¢/ton used to pay principal and interest on bonds sold to construct bulk loading facilities. Excess is rebated. 20¢/ton allocated to local government by set formula.
Alaska			
Oil	Production	The greater of 12.5% of gross value, 60¢/bond of old oil or 80¢/bond of new oil if API gravity ≥ 27 degrees	Rate is increased (decreased) .005¢ for each degree variation in API (up to maximum of 40 degrees). Rate is then multiplied by a production factor. Per barrel rate may be changed if price changes. Paid in lieu of property and other output-related taxes.
Gas	Production	The greater of 10% of gross value or 6.4%/1,000 cubic feet	Rate is multiplied by a production factor. Paid in lieu of property and other output-related taxes. Rate may be changed if price changes.

[a]Gross value is defined as price times quantity. No property taxes are included even though they may be output-related. See table 1-3.

Table 1-1 continued

State/Mineral	Name	Base and Rate	Remarks
Arizona			
All minerals	Occupational gross income	2.5% of gross value	Price determined by reference to market publications less out-of-state processing charges and transport costs. 1% of the 2.5% is allocated by set formula. From 6/1/78 to 6/30/80 rate on copper mining and smelting is 2%.
Arkansas			
Oil	Severance	5% of gross value if well produces more than 10 barrels/day. 4% of gross value if well produces less than 10 barrels/day	
	Special fund	5 mills per barrel	Used to establish the Arkansas Oil Museum.
Gas	Severance	0.3%/1,000 cubic feet	
Oil and gas	Conversion	10 mills/barrel or 1 mill/1,000 cubic feet	
Iron ore	Severance	2¢/ton	
Barite, bauxite, titanium ore, manganese and manganiferous ores, zinc ore, cinnabar, and lead ore	Severance	15¢/ton	All severance taxes are allocated between state and local governments by set formula.

Gypsum (sold for out-of-state use), chemical-grade limestone, silica sand, and dimension stone	Severance	1.5¢/ton	
Crushed stone (most varieties)	Severance	1¢/ton	
Others (including diamonds, salt, and so on)	Severance	5% gross value	
California			
Oil and gas	Severance	Rate varies yearly, 0.02%/barrel in 1977	Considered a regulatory tax.
Colorado			
Metallic minerals	Severance	2.25% of gross revenue in excess of $11 million	Property taxes are allowed as a credit up to 50% of the severance tax due. Funds go into state trust.
Molybdenum	Severance	15¢/ton	
Oil and gas	Severance	2% if gross revenue less than $25,000; 3% $25,000 to 99,999; 4% $100,000 to 299,000; 5% over $300,000	87.5% of property taxes may be used as a credit.
Coal	Severance	60¢ in open pit; 30¢ underground plus 1% increase for every 3% increase of wholesale price index	First 8,000 tons per quarter exempted. Lignite gets additional 50% credit.
Oil shale	Severance	maximum 4% gross value of the 4th year of operation	25% credit for on-site methods.

Table 1-1 continued

State/Mineral	Name	Base and Rate	Remarks
Florida			
Oil	Severance	8% gross value (if average production is greater than 100 barrels/day)	7/8 to state general fund; 1/8 to county where produced.
Gas	Severance	5% gross value	80% to state general fund; 20% to county where produced.
Solid minerals	Severance	5% gross value	75% to general fund; 25% to land reclamation trust; credit for property taxes up to 20% of taxes due. Further credits if firm has own reclamation program.
Phosphates	Severance	10% gross value	50% to general fund; 50% to land reclamation trust.
Idaho	None	All are income-based	
Kentucky			
Coal	Severance	4.5% gross value	Minimum 50¢/ton.
Oil	Severance	0.5% gross value	Up to 1% additional may be collected by counties.
Louisiana			
Sulfur	Severance	$1.03/long ton	All severance taxes distributed by set formula between states and localities.

Mineral	Tax	Rate	Notes
Salt	Severance	6¢/ton	
Sand and gravel	Severance	3¢/ton	
Marble	Severance	20¢/ton	
Coal	Severance	10¢/ton	
Oil	Severance	12.5% gross value	Rate can be reduced for small wells and special circumstances.
Gas	Severance	7¢/1,000 cubic feet	
Michigan			
Oil and gas	Production	2% gross value	Paid in lieu of property taxes.
Minnesota			
Semitaconite concentrate	Severance	10¢/ton plus 1/10 of 1% of grade greater than 55%	5¢/ton base if processed in state.
Copper-nickel	Severance	1% of value plus 2½¢/ton plus 10% of bases for each 0.1% increase of valuable content above 62%	
Tacomite, iron sulfide concentrates	Occupation	$1.25/ton plus 1.6% for each 1% valuable content above 62%	Rate increases with steel-mill products index.
Mississippi			
Oil and gas	Severance	6%/barrel or 6% value 3 mills/cubic foot or 6% value	Paid in lieu of property taxes. Revenue distributed to state and local governments by set formula. Slight tax for administration of conservation laws.
Other minerals	Occupation	5% gross value	

Table 1-1 continued

State/Mineral	Name	Base and Rate	Remarks
Montana			
Cement and gypsum	License	22c/ton of cement 5¢/ton of gypsum	
Coal	Severance		G.V. = gross value. First 20,000 tons exempt. Distributed to various counties and state agencies by set formula.

Btu/lb	Surface	Underground
> 7,000	12¢/ton or 20% G.V.	5¢/ton or 3% G.V.
7,001 to 8,000	22¢/ton or 30% G.V.	8¢/ton or 4% G.V.
8,001 to 9,000	34¢/ton or 30% G.V.	10¢/ton or 4% G.V.
90,017	40¢/ton or 30% G.V.	12¢/ton or 4% G.V.

State/Mineral	Name	Base and Rate	Remarks
Metalliferous ores	License		

Gross Value	Rate
0–100,000	0.15 %
100,001–250,000	0.575%
250,001–400,000	0.86 %
400,001–500,000	1.15 %
500,000 and over	1.438%

State/Mineral	Name	Base and Rate	Remarks
Micaceous ores	License	5¢/ton	
Oil	Severance	2.1% first $6,000 of gross value 2.65% excess over $6,000 of gross value	
All minerals	Resource trust	$25 plus 0.5% gross value in excess of $5,000	Used to rectify environmental damage.

Nebraska

| Oil and gas | Severance | 2% of gross value |
| | Conservation | 4 mills/dollar |

Nevada

| Oil and gas | Conservation | 5 mills/barrel of oil or/50,000 cubic feet of gas |

New Mexico

Uranium	Severance		
		Value/lb	Tax (marginal rate)
		0–5.00	1 %
		5–7.50	1.6 %
		7.50–10.00	2.0 %
		10.00–15.00	3.0 %
		15.00–20.00	4.0 %
		20.00–25.00	5.0 %
		25.00–30.00	7.0 %
		30.00–40.00	9.0 %
		40.00–50.00	12.5 %
		750.00	$3.24

			If contracts in force prior to 1977, tax is 1.25%.
Oil and gas	Severance	.45¢/barrel or 5¢/1,000 cubic feet plus increase for change in CPI	
Coal	Severance	38¢/ton plus increase for change in CPI (18¢/ton steam coal)	Funds from severance taxes go to state.
Potash	Severance	2.5% gross value * 1/3	
Copper	Severance	0.5% gross sales * 1/2	
Others	Severance	0.125% gross sales * 1/2	
Hard minerals	Resource Excise	0.75% gross value	

Table 1-1 continued

State/Mineral	Name	Base and Rate	Remarks
North Dakota			
Oil and gas	Production	5% gross value	Paid in lieu of property taxes. Revenues divided between state and localities by set formula.
Coal	Severance	65¢/ton plus 1% for every 3% change in WPI	Not in lieu of property taxes. Revenues divided between state and localities by set formula.
Ohio			
Coal	Severance	4¢/ton	Severance tax revenue goes into state general fund.
Salt	Severance	4¢/ton	
Limestone and dolomite	Severance	1¢/ton	
Oil	Severance	3¢/barrel	
Gas	Severance	1¢/1,000 cubic feet	
Oklahoma			
Oil and natural gas	Production	7% of gross value	Paid in lieu of all other taxes.
Uranium	Production	5% of gross value	
Other minerals	Production	0.75% of gross value	
South Dakota			
Oil and gas	Severance	3% of gross value	Paid in lieu of license tax.

Tennessee			
Oil	Severance	5¢/barrel	Revenue allocated to various agencies by set formula.
Gas	Severance	5% gross value	
Sulfur	Severance	$1.03/long ton	
Utah			
Metals	Occupation	1% gross value	First $50,000 is exempt. Portion of tax must be prepaid.
Oil and gas	Occupation	2% gross value	
West Virginia			
Coal	Occupation	3.85% gross value	
Limestone or sandstone	Occupation	2.2% gross value	
Oil	Occupation	4.34% gross value	
Gas	Occupation	8.63% in excess of $5,000 of gross value	
Wyoming			
All minerals	Mining excise	2% gross value	Revenue to state general fund.
Coal, uranium oil, gas	Severance	2% gross value	Revenue to mineral trust fund.
Coal, uranium	Severance	1.5% gross value	Revenue to capital facilities account.
Coal	Severance	5% gross value	Revenues to various funds.

Sources: Written correspondence from state tax administrators; Stinson (1978); Gillis (1979); State tax statutes; Commerce Clearing House: State Tax Guide; Yasnowsky and Graham (1976); and Steering Committee on the Impact of Taxation on Energy Markets, National Academy of Sciences (1979).

Note: This tax table was compiled using information from a variety of sources. The information was cross-checked as far as possible to ensure consistency and to include the most recent tax laws.

a secure revenue base is also reflected in the linking of tax rates to inflation indicators (South Dakota) and in the calculation of taxes using both a fixed dollar per ton rate and an ad valorem rate with collection of the larger amount (Montana and Alaska). Finally, tax revenues are used for a variety of purposes. In Alabama, output taxes go directly to the general fund and are paid in lieu of property taxes, while in North Dakota these taxes are earmarked for various categories of expenditure at both the state and local levels.

Profits Taxes

Profits taxes, if imposed in a state, are paid by every corporate entity. So in the absence of special tax privileges granted to any industry, mining enterprises are treated equitably under this tax. The profits tax has an advantage over other types of taxes in that it considers both the costs of operation and the depletion of the resource base. Taxes are paid only when revenues exceed costs, which again implies that taxes are levied on a more equitable basis. Also because costs are considered, the tax recognizes the ability to pay. Thus there is no inherent bias against low-profit, small marginal mines.

From the state's perspective there are several difficulties with employing the profits tax. One is the cost of administration. In order for the tax to be applied equitably, accurate books and procedures must be maintained. States have partly attacked this problem by merely adjusting the tax base used for federal tax purposes (table 1-2). In effect, some states piggyback the federal taxes, substantially reducing administration at the state level. However, this advantage is partly offset by revenue fluctuations induced by changes in the federal tax laws.

Second, there is the problem of allocating the income of a corporation that operates in more than one state. The major allocation rule for income is known as the "ABC" rule, which allocates income to the state on the basis of three ratios: the ratio of total sales in the state to total corporate sales; the ratio of total assets in the state to total corporate assets; and the ratio of total employment in the state to total corporate employment. This is an arbitrary procedure that will typically not reflect the profitability of a mine in a given state.

Third, there is the intrafirm pricing problem for integrated firms.

Interstate transactions may be of an intrafirm nature, with no market price established for output. This is a severe problem in the mining sector where quality variations are reflected in the market price but may be difficult to detect on an intrafirm transaction.

Finally, there is the issue of allocating corporate overhead and compensation of corporate officers. While these are legitimate expenses at the corporate level, they are difficult to prorate to different state operations.

Table 1–2 presents a summary of profits taxes in the states surveyed. As noted, most states use the federal tax base with only minor adjustments. Those that do not (Arizona, Minnesota, and Michigan) use different methods in computing loss carry-forwards, depreciation, and expensing. The rates differ from a low of 2.35 percent (Michigan) to a high of 12 percent (Minnesota). Also some states allow a deduction for federal taxes (Alabama) while others do not (Arizona). Finally, some states add back other state taxes paid to the federal government while others do not. Those that do (Colorado, for example) increase the net effective rate of these taxes on the mine's operation and so increase the burden of taxation on the mine.

Property Taxes

Ad valorem property taxes are the oldest form of taxation imposed on the mining sector. Ideally the tax should be a tax on the "wealth" of the mine. However, the wealth of a mine is very difficult to estimate. Neither future prices and costs nor the geological characteristics of the deposits are known with certainty. Furthermore, local tax assessors are not usually trained in the techniques necessary to estimate these parameters. To compound these difficulties, as will be shown, property taxes induce the firm to extract the ore at a faster rate than would be the case in the absence of the tax.

As shown in table 1–3, several states have realized these difficulties and in effect use other forms of taxes in lieu of property taxes. Only Arizona attempts to estimate the present value of the operation. Other states employ gross-proceeds methods, output taxes, or some form of net-proceeds tax. Net-proceeds taxes are similar to profits taxes in that they aim to measure the net profits from an operation. The major distinctions between net-proceeds and corpo-

Table 1-2
Profits Taxes

State	Rate	Federal Income Tax Deductible	Federal Income Used as Base	Allow Federal Depletion	Remarks
Alabama	5%	Yes	No	No	
Alaska	9.4%	No	Yes	Yes	Current expenses of unsuccessful exploration.
Arizona	10.5% income > 6,000	Yes	No	No	Some expenses for exploration.
Arkansas	6% income > 25,000	No	No	No	
California	9%	No	No	No	Minimum tax = $200 (reduced to $25 for gold). Current expensing of exploration and development. Percentage depletion allowed.
Colorado	5%	No	Yes	Yes (except shale)	
Florida	5%	No	Yes	Yes	
Idaho	6.5%	No	Yes	Yes	
Illinois	4.0%	No	Yes	Yes	
Kansas	6.75% income > 25,000	No	Yes	Yes	Accelerated deduction for oil cost.
Kentucky	5.8% income > 25,000	No	Yes	Yes	
Louisiana	8% income > 200,000	Yes	No	Yes	Depletion: oil = 38%; coal = 15%; and sulfur = 23%.
Michigan	2.35%	No	Yes	Yes	
Minnesota	12%	No	No	No	Minimum tax = $100; federal depletion allowed for copper and nickel.

State	Rate				Notes
Montana	6.75%	No	Yes	Yes	ABC rule at firm's option.
New Mexico	5%	No	Yes	Yes	
North Dakota	6% income > 15,000	Yes	Yes	Yes	
Oklahoma	4%	No	Yes	Yes	Oil depletion.
Pennsylvania	9.5%	No	Yes	Yes	
South Dakota	5.5%				
Utah	4%	No	No	No	Minimum = $25.
West Virginia	6%	No	Yes	Yes	

Sources: Written correspondence from state tax administrators; Stinson (1978); Gillis (1979); State tax statutes; Commerce Clearing House; State Tax Guide; Yasnowsky and Graham (1976); and Steering Committee on the Impact of Taxation on Energy Markets, National Academy of Sciences (1979).

Note: This tax table was compiled using information from a variety of sources. The information was cross-checked as far as possible to ensure consistency and to include the most recent tax laws.

rate-profits taxes are the use of cost depletion and straight-line depreciation; the disallowance of corporate overhead, research and development, and executive compensation; and differences in inter-state profit allocation rules. In effect, net proceeds are closer to the economic definition of profit from the mine than is the measure for corporate-income taxes.

The use of other types of taxes in lieu of ad valorem property taxes offers other advantages. First, the tax is usually a statewide tax, thus ensuring equity among the state's mineral producers. Second, administration is transferred from the local level to the state. The state's larger and more highly trained bureaucracy provides scale economies in administration and more effective assessment, collection, and enforcement. Finally, the funds are allocated to the localities in a fashion that attempts to measure the costs of local services attributable to the mine. Mines in one county may use public services from another county: in particular, schools, electricity, and water. Under the old system, the counties providing the services could not obtain reimbursement when the mines were technically in a different jurisdiction.

In summary, states have generally moved away from the old types of property-tax administration. The alternatives also present both administrative difficulties and adverse economic incentives, but these drawbacks have been judged lesser than those of property-tax administration.

Table 1-3
Property Taxes

State/Mineral	Base	Remarks
Alabama	None	Production tax paid in lieu of property tax.
Alaska	None	Other taxes paid in lieu of property tax.
Arizona		
Oil and gas	Gross value of production	
Other mines	60% of full coal value	Present value of planned extraction using five-year average profit margin.

Table 1–3 continued

State/Mineral	Base	Remarks
Arkansas	Local assessment	
California	Local assessment	
Colorado	Gross value	
Florida	Local assessment	
Kentucky	10¢ per $100 of assessed value	State rate; local rates vary.
Louisiana	None	
Michigan	2% of five-year average net-production value	
Mississippi	None	Other taxes replace property tax.
Montana	Gross value	
Nevada	Net proceeds	
New Mexico		
Uranium	25% gross value less royalties	
Oil and gas	9% gross value	
Coal and other minerals (productive)	Profits	300% of value of profits times 1/3 which is statewide assessment ratio.
Nonproductive properties	Assessed value	
North Dakota	None	Severance taxes paid in lieu of property taxes.
Oklahoma	None	Severance taxes paid in lieu of property taxes.
Oregon	Local assessment	
Pennsylvania	Local assessment	
Utah	2 times average of net proceeds for 3 years non-metalliferous 30% capitalized net income for 5 years plus $5/acre	New mines must prepay taxes to offset increased demands on local public services.
West Virginia	Local assessment	
Wisconsin	Net proceeds	Progressive rates.
Wyoming	100% gross proceeds	

Sources: Written correspondence from state tax administrators; Stinson (1978); Gillis (1979); State tax statutes; Commerce Clearing House: State Tax Guide; Yasnowsky and Graham (1976); and Steering Committee on the Impact of Taxation on Energy Markets, National Academy of Sciences (1979).

Note: This tax table was compiled using information from a variety of sources. The information was cross-checked as far as possible to ensure consistency and to include the most recent tax laws.

Discussion

The preceding discussion shows the wide variety of tax methods used by states. There are numerous reasons for this variance. First, the size of the state's economic base (and thus the tax base) derived from the mining sector is an important factor. States that have relatively large mining sectors must rely more heavily on mineral taxation to support state and local services. Presumably, this leads to a more diverse and sophisticated set of tax policies, since these states need to collect relatively more revenue from the mining sector while ensuring the continued economic viability of their resource base. States that depend on the mining sector for a substantial part of their revenues also employ and train tax administrators to deal exclusively with the mining sector. Arizona is a prime example: the state assesses all mineral property, has its own mineral tax division, trains assessors, and expends considerable effort keeping up to date on cost, market, and technological trends.

Second, state legislatures differ in their perceptions of the role of mineral taxation. Some regard these taxes as a means of collecting resource rents and a method of tax exportation, while others view mines as corporate enterprises and attempt to treat them as such.

A third reason is the attitude of some states that state taxes are small relative to federal taxes, and that, since state taxes are deductible from federal taxes, the impact of state taxes on the economic behavior of a mine is small. This may have been true in the past but chapter 4 shows that it is no longer the case. The appendix provides data on the size of the revenues generated by these taxes.

2 Mining Decisions

The objective of the mining firm is to profitably extract and process the valuable contents of a mineral deposit. This objective is achieved, if at all, only after a long and costly process. This chapter describes the process and models it from an economic perspective. Proper recognition of the actual structure of mining decisions, and of the influence of geological factors, is essential for prediction of responses to changes in economic conditions.

Description of the Mining Process

Figure 2-1 presents schematically the major phases of a typical mining cycle, with estimates of the time and cost necessary to complete each phase. In practice, the time and cost requirements can vary substantially across minerals and across mines of the same mineral, on account of geological differences, location, and other factors. The numbers on figure 2-1 should therefore be considered as only representative. Each phase is now described in some detail.[1]

Exploration

A typical exploration program usually involves a three-step procedure: preliminary exploration and sampling; detailed exploration; and estimation of reserves and recovery factors.[2] The first step aims to determine areas of possible mineralization or anomalies over large regions. If there has been previous mining activity in a region, old mappings, surveys, and other historical documents are studied. Geological maps and other material compiled by state or federal agencies are also used. Specially equipped airplanes (or even satellites) are then used to photograph the region and to conduct magnetic and other geographical surveys.[3]

Activity	Time to Complete Activity	Cost
Exploration	2–4 years	$10 million
Investment	2–4 years	$50 million
Development	2–4 years	$100 million
Exploitation	10–20 years	
Profit		

Source: Adapted from Thomas (1973), p. 60.

Figure 2-1. Cycle of Mining Operations

The results of this work identify promising anomalies which are studied in more detail in a series of preliminary on-site inspections. These inspections consist of general geophysical or geochemical tests on chip samples from outcroppings or streambeds. These tests, along with detailed mappings, enable a determination of the areas which hold the greatest prospects for mineralization and thus merit more intensive exploration.[4]

The methods employed in the detailed exploration phase depend on the type of mineralization and the geological characteristics of the particular deposit. Costeaning and shallow drilling are used for deposits with only a thin layer of overburden. In cases where the mineralized area is farther below the surface, diamond drilling is usually employed. Regardless of the technique, the objective of the drilling is to determine as accurately as possible the size of the deposit, the metallic content, the fault structure, and other geological, geophysical, and geochemical features of the deposit. This process involves taking samples of the area at various points and depths in order to map the extent of the deposit.[5] The samples are also used to determine what processes will be necessary to recover the valuable material from the waste rock. The types of information obtained include (1) length, width, and depth of mineralized area; (2) fault structure and other discontinuities; (3) specific gravities; (4) moisture levels; (5) grade of ore; and (6) nature of overburden.

This information from the samples is used to place the reserves into three broad categories.

1. Proved reserves: ores that have been both delineated and measured; tonnage and grades are known within 5-percent error.
2. Probable reserves: tonnage and grades are computed from widely spaced samples and from geological projections; errors in estimates are usually less than 20 percent.
3. Possible ore: no samples available; estimates based on inferences solely on geological structure and geophysical anomalies.[6]

The estimates of reserves and other factors are then used to compute an estimate of the total tonnage of ore contained in the deposit, that is, theoretical reserves. This estimate is adjusted for the fault structure and other geological factors to obtain an estimate of how much ore can be extracted, that is, minable reserves.[7]

Once the sample information has been analyzed, a determination is made of alternative methods which might be used to extract and process the reserves. This involves choosing a set of mining methods (open pit, underground) and a concentrating process (flotation, separation) which are appropriate given the available geological information. For each mining method, the mineral reserve estimates are modified in two ways. The mining process itself may cause some ore to be left behind. This results from the necessity of roof and floor supports in underground methods, or overburden removal in open-pit methods. To allow for this, minable reserves are adjusted by an estimate of the extraction ratio (the proportion of ore left unmined) to derive net recoverable reserves.[8]

Dilution of ore also occurs in the mining process. It is inevitable that some waste rock will get mixed with ore. This reduces the value of ore mined and increases the tonnage of ore which must be handled (run-of-mine ore) to recover the same amount of valuable product. Therefore, an adjustment called the "dilution factor" must be made to account for the increased tonnage which must be processed.

The final adjustment is made for losses incurred in the separation or concentrating process. In most cases, raw ore is processed at or near the mine site to separate the valuable material from the waste. In the case of metal mining the ore is crushed and a chemical or magnetic process used to separate the metal from the other elements. The method used in the process is determined, in part, by the characteristics of the ore. Some valuable material will be lost as a result of the separation process and so a further adjustment must be made in order to estimate the tonnage of material which will be sold. This adjustment in known as the "recovery rate."

To summarize, the exploration process is a method where various areas are selected for intensive examination to determine the nature of reserves. In addition to estimating the reserves in the deposits, various adjustments are made to account for losses in the mining and recovery process itself. The results of this stage include estimates of recoverable ore and material available for sale, in addition to reserve estimates.[9]

Investment

If exploration has been successful, a determination of the mine's profitability must be made. This determination is usually a two-step procedure.[10] A "quick" assessment of the mine's profitability is made to decide if the expense and time necessary to complete detailed feasibility studies are warranted. This assessment is based on current price projections, the average grade of ore, and crude cost estimates.

If the chances for profitable operation are promising, detailed studies are initiated. One part of this process is a series of engineering studies which determine the best technology, scale, and processing techniques, under alternative methods of operation.[11] The engineering detail is quite specific in order to determine capital and labor requirements and costs, as well as ordering the reserves for production. Marketing studies are made to determine price projections and access to markets. Preliminary negotiations with potential buyers are initiated to determine quality and quantity characteristics which form the basis for long-term contracts. Finally, alternative methods of financing are evaluated and banks are contacted to obtain loans.[12]

The information from the engineering and marketing studies is then combined to select the most profitable method of operation. The procedure commonly used is to determine the profits from an annual extraction and development profile for a given scale of operations and a specific technology. The size of the operation is changed (in increments of about 50 percent) to generate new estimates of profitability. Finally, sensitivity studies are made to determine the effect of changing conditions on profitability under alternative mining methods. The final decision on the type and scale of the mine to be constructed is based on a number of criteria. First, the present value is calculated using discount factors that reflect attitudes toward risk. The internal rate of return is calculated for comparison with other investments. Minimum returns may range from 12 to 25 percent. Finally, the undiscounted payback period is calculated to estimate the time necessary to recapture the initial investment. Based on these and other criteria, the technology and plant size are chosen which yield the greatest estimated present value of profits.

Once financial arrangements are made the construction and development stage is begun. Before any ore is extracted and processed, the mine, processing plant, and storage and transportation facilities must be constructed. Not all of the mine is constructed at once, however. Rather, development expenditures of different parts of the mine are sequenced in a manner consistent with the long-range development profile. Even if it were technically feasible to develop the entire mine at once, economic considerations would dictate a sequence of incremental expenditures. If too many areas are developed relative to capacity, additional investment costs would be incurred early in the operation which have no immediate payback. In addition, overall costs would be increased as a result of keeping developed areas not currently in operation in safe and productive condition. Thus it is better to sequence mine development so that some areas are developed as others are exhausted, to lower overhead and defer incremental expansion costs until conditions warrant them.

Extraction

After all this time and expense involved in exploring and constructing the mine, extraction of ore can proceed and a positive cash flow generated. At this stage the firm is constrained by all its previous decisions. Capacity and technology are in place, limiting the amount of ore that can be currently extracted and determining both overhead and variable costs. The only variables which can be controlled at this level are the quantity of ore processed, and its quality. The cash flow generated from current operations is used to repay debt, pay returns to equity, begin operations and exploration activities in other mines, or reinvest in the current operation.

Although the firm is constrained by all its previous decisions, the extraction process is still dynamic in nature. Production schedules must be established for a six-to-twelve-month period and must be reevaluated in light of new developments and changing market conditions.

The Effects of Uncertainty

Uncertainty in mining operations arises from three sources. First, the tonnage and quality of a deposit are not known with certainty. Until shafts are developed (or overburden removed) the exact delineation of particular areas is uncertain. In addition, the tonnage and quality of final output may not be known until the ore is extracted and processed. Second, the future economic conditions (in particular, prices and costs) are unknown and cannot be perfectly predicted. Third, there are social and political uncertainties which may be beyond the control of the operator. Governments can change tax and environmental policies at some future date, thereby affecting the profitability of the operation.[13]

These risks are compounded by the time lag between exploration and extraction, and by the costs of obtaining and processing information. An investment decision made today will not generate sales until three to five years hence. This means that funds must be committed for a substantial period of time before any profits are forthcoming. Furthermore, if an attempt were made to account for every possible eventuality, the firm would spend too much money and time to warrant an investment. Therefore, trade-offs must be made between accuracy of prediction and the risks assumed by the operators.

The mining process follows a logical sequence from exploration to exploitation and thus it is natural to segment the decisions according to each phase. This segmentation allows each decision to be considered separately; for example, should there be more exploration; how much development is warranted; and how much ore should be extracted and processed this month?[14] Such a procedure does not eliminate risks, but it allows incorporation of new information and reevaluation of options based on previous results.

Since not all decisions are made and reevaluated at once, simplifications or rules of thumb have to be used. These simplifications vary from one decision to another. They relate to (1) length of the planning horizon; (2) areas of the mine under consideration; and (3) averages (of various magnitudes) of such variables as costs, days worked, and tonnages of ore processed.[15] For instance, exploration does not provide complete information about the nature of reserves. Projections must be made, based on samples and past experience. The investment decision is made on the basis of these estimates and annual averages of prices and costs. At the current extraction stage, currently developed areas are the only consideration and the planning period may be less than a year.

This segmentation of the process results in a hierarchical decision structure which allows for revisions as time proceeds. This procedure also enables variables which must be determined at one level to be treated as exogenous at another. For example, exploration provides reserve estimates which are used in the investment decision. The investment determines the technology and capacity, which in turn are treated as given when extraction plans are made. Further, information acquired at one level can be used in the reevaluation of plans at other levels. For example, information gained from current extraction, such as profits and revised ore estimates, is used in evaluating possible changes in capacity and modifications to development and future exploration activity.

One consequence of this decision structure is that uncertainty is reduced at lower levels. At the current extraction level, the manager cares only about short-term variations in prices and costs. Prices ten years from now are of no concern. The areas of possible extraction have been developed and the properties of the ore are known. Finally, capacity is in place and there is a history of operations to aid in reevaluation. The initial investment decision, in contrast, must be made without the benefit of this information. Prices and costs must be projected far into the future so that annual averages must be used to make any projections at all. The structure of the ore body is an estimate and the technological implications of alternative investment strategies are unknown.[16]

A further consideration of practical importance is that modifications to long-run decisions take time and incur adjustment costs. For instance, if prices are higher than expected, a larger capacity and more development may be needed to exploit this trend. But an instantaneous adjustment is not possible and there is always the risk that, by the time any change has been effected, the interim evolution of events will have rendered it unprofitable.

Effect of Risk: An Example

For an illustration of how uncertainty can affect the decisions made in mining operations, consider the hypothetical tin deposit described in table 2-1. The price of tin for 1960-1972 is also reported in the table. If the operator knows all geological and economic factors with certainty, the only problem is to choose an investment which will yield the greatest present value to the firm. Such an investment and the corresponding optimal extraction profile are reported in table 2-2. The calculations are based on an investment cost of $200 million, extraction cost of $10 per ton of ore, a capacity of 2 million tons of ore per year, and a 10-percent discount rate.

Two properties of this plan should be noted. First, the ordering of the grade selection profile corresponds to the ranking of discounted prices (see column 3 of table 2-1). The best two grades of ore are extracted in 1965 when the price discounted at 10 percent is the highest. The reason for this behavior is that more metal per ton is obtained from high-grade ore than from low-grade ore and thus current revenues are always maximized by extracting the best ores. However, the best grades of ore are in limited supply and so should be extracted when their contribution to profitability is the greatest; that is, in the periods with the highest discounted prices.[17]

Second, the mine closes in 1971 even though there are 16 million tons of reserves left in the deposit, because the remaining reserves cannot be profitably extracted; that is, they are below the cutoff grade. For instance, if the 25-percent ore were extracted in 1972, a $2.25 million loss would result. Therefore, even in the absence of uncertainty, it may be optimal from the operator's perspective to leave some ore behind.[18]

When uncertainty is incorporated into the decision structure the planned operation may change substantially. For simplicity, it will be assumed that the characteristics of the reserves are known, so that

Table 2–1
Annual Price of Tin (1960–1972): Hypothetical Tin Deposit

Year	Price (¢/lb)	Price Discounted @ 10%	Ranking of Discounted Prices
		Annual Price of Tin	
1960	101.40	92.18	4
1961	113.27	93.61	3
1962	114.61	86.11	5
1963	116.64	79.67	7
1964	157.72	97.93	2
1965	178.17	100.57	1
1966	164.02	84.17	6
1967	153.40	71.56	8
1968	148.11	62.81	10
1969	164.43	63.39	9
1970	174.13	61.03	11
1971	174.36	55.56	12
1972	177.47	51.41	13

Grade (%)	Tons (millions)	Grade (%)	Tons (millions)
		Hypothetical Tin Deposit	
2.80	1	0.70	2
2.60	1	0.65	2
2.40	1	0.60	2
2.20	1	0.50	2
2.00	1	0.40	4
1.60	1	0.30	4
1.20	1	0.25	8
0.80	1	0.20	8

Source: American Metal Market (1978), p. 243.

uncertainty arises only because of lack of information about future prices and costs. One procedure commonly used to project prices and costs is extrapolation along a linear trend.[19] Suppose that an economic analysis of historical data reveals that prices and costs rise at a 5-percent annual rate. This means that the decision maker will take $101.40 per ton and $10 per ton of ore as base-period prices and increase each by 5 percent a year through the planning period.

Table 2–2
Optimal Extraction Profile for Tin Deposit

Year	Average Grade Extracted (%)	Undiscounted Profit (million $)	Discounted Profit (million $)
1959		− 200.00	− 200.00
1960	1.00	20.56	18.69
1961	1.80	61.55	50.87
1962	0.70	12.09	9.08
1963	0.60	7.99	5.46
1964	2.30	125.10	77.68
1965	2.70	172.42	97.33
1966	0.65	22.65	11.62
1967	0.50	10.68	4.98
1968	0.40	3.70	1.57
1969	0.40	6.31	2.43
1970	0.30	0.90	0.31
1971	0.30	0.92	0.29
Total		244.87	80.31

Source: Conrad (1978*b*)
Marginal cost: $10/ton.
Capacity: 2 million tons/year.
Discount rate: 10%.
Investment cost: $200 million.

Another method used to account for uncertainty is to increase the discount rate.[20] In the present example a 15-percent discount rate will be used, representing a 50-percent premium over the rate in the absence of uncertainty. Finally, operators often assume that the average grade of ore will be extracted in each year. Since prices are not known, an optimal grade-selection profile cannot be determined. The average-grade rule is used to simplify the calculations and to lower the range of possible extraction profiles, in order to keep down computational costs. When a detailed plan is made, the firm may alter this assumption.

The net effect of these three adjustments is to lower the present value of the investment. This is clear from an examination of table 2–3, where the present value is negative. Note also that the planned mine life has been reduced to eight years. This is a consequence of the method used to project and discount prices and costs. In effect, the cutoff grade is the one which would be calculated using the base-

Table 2–3
Extraction Profile Using Average-Grade Rule

Year	Average Grade Extracted	Undiscounted Profit	Discounted Profit
1959		− 200.00	− 200.00
1960	1.28	33.50	29.13
1961	1.28	35.17	26.59
1962	1.28	36.93	24.28
1963	1.28	38.77	22.17
1964	1.28	40.71	20.24
1965	1.28	42.75	18.48
1966	1.28	44.89	16.87
1967	1.28	47.13	15.41
Totals		119.85	− 26.83

Undiscounted payback period: 5.35 years. Present value using Hoskold method: − 36.15 (calculated using average annual earnings, an 8-percent safe rate, and a 15-percent risk rate).

year price and cost estimates, thereby forcing more ore into the sub-marginal category.

The higher discount rate further reduces the present value of future earnings from the planned extraction. Although undiscounted annual profits consistently rise (because of the method of price and cost extrapolation), their present value constantly falls. Finally, the average-grade extraction rule further reduces the estimate of present value because it does not allow the matching of higher quality of ore with higher discounted price. (In the present example, the best ore should be extracted in 1965.)

The fact that this scale of operation produces a loss does not necessarily mean that the mine will not open. There may be another plant size which can generate a positive present value. Such a plant size and corresponding optimal extraction profile are presented in table 2–4. Note that a lower capacity generates a higher cost per ton of ore extracted. This means that less ore is planned for extraction but its average grade is higher (that is, the cutoff grade is higher). Note also that the planned life of the mine is longer than with a large investment, because capacity is smaller relative to economically recoverable reserves. The mine manager may further restrict the decision by arbitrarily imposing a fixed planning horizon.[21]

Table 2–4
Optimal Extraction Profile with Lower Capacity

Year	Average Grade Extracted (%)	Undiscounted Profit	Discounted Profit
1959		−25.00	−25.00
1960	2.4	4.56	3.96
1961	2.4	4.78	3.62
1962	2.4	5.02	3.30
1963	2.4	5.28	3.02
1964	2.4	5.54	2.75
1965	2.4	5.82	2.51
1966	2.4	6.11	2.30
1967	2.4	6.41	2.10
1968	2.4	6.73	1.91
1969	2.4	7.07	1.75
Totals		32.32	2.22

Undiscounted payback period: 4.97 years. Present value using Hoskold method: 1.17.

These examples highlight several basic points about the effects of uncertainty on the operations of the mine. First, investment and extraction are not all-or-nothing decisions. The optimality of plant size is relative to given economic and geological conditions. In general, uncertainty reduces the amount the firm will be willing to invest at the beginning of the operation. This results in higher extraction costs, lower total recovery and, in most cases, a shorter planned life.

Second, the acquisition of information over time may suggest a subsequent reevaluation of the plan. The examples cited are limited to what the firm initially plans to extract. Even with a given investment it is possible that the actual behavior of the firm will be significantly different once new information is available. If prices rise (fall) relative to costs, more (less) ore may be recovered and the life of the mine may be longer (shorter) than anticipated. However, the firm must bear the costs of its past decisions in making these adjustments. This by no means implies that the operator's behavior is not systematic. Rather, uncertainty complicates the interpretation of economic behavior.

Summary

The preceding description of the mining-decision process was intended to highlight the economic and geological variables which influence the behavior of a mine operator. An understanding of this decision structure is important in determining the effects of public policy, particularly taxation, on the overall development of the mining sector in a particular state. At different stages of the mining cycle the firm is operating on different cost curves and using different assumptions to make decisions. Therefore, a tax imposed on a mine that is fully developed may have different effects than the same tax on a mine which has yet to be completed. Also a tax may have no effect on the extraction profile once the investment is made, yet might change the future development of the mine.

A second important aspect of this decision structure is the joint influence of geology and economics. Each mine is unique and, even if economic conditions are the same for all, decisions will typically vary across mines. Accordingly, a public-policy initiative, such as a tax, may produce different responses. Geological factors cannot be overlooked.

Finally, it should be emphasized that the decision structure used in mining is not peculiar to this industry. All economic agents must confront uncertainty and limited computational capacity, and this is reflected in planning structures.[22] Individuals plan for retirement and plan next month's budget. The decisions are clearly interrelated but they are not made simultaneously. There are also analogs to the exploration and the long lead times necessary to obtain a positive return that characterize mining. For instance, the drug industry spends large sums on research (exploration) for new drugs, much of which is spent on abandoned projects (dry holes). This research is conducted with the knowledge that a competitor may make the discovery first and control the market (if a cure is not found in the interim which makes the drug obsolete). Finally, the time from initial discovery to positive cash flow is on average quite similar to lead times in mining. So while risks exist at all levels of the mining problem, they must be taken in context and evaluated relative to risks inherent in other economic activity before appropriate policy can be formulated.

Notes

1. This section is only an introduction to the complexities of the mining process. For a more complete description see McKinstry (1948), Truscott (1962), Pfluder (1968), Just (1976), and Thomas (1973).

2. See Thomas (1973) and Allais (1957) for details.

3. Byrne and Sparvero (1969) and Maximov et al. (1973) describe such processes.

4. See Preston (1960) and Patterson (1959) for a complete description.

5. The determination of where to drill may be based on mathematical methods, such as those described by Agterberg and Kelly (1971). For a description of diamond drilling see Koch and Link (1971).

6. Thomas (1973).

7. See Patterson (1959) for a description of this.

8. Hazen (1967), Koch (1971), and Link et al. (1971) describe such factors.

9. See Cooper, Davidson, and Reim (1973) for a discussion of the trade-offs between accuracy and cost.

10. Details of this process are found in Lewis and Bhappu (1975), Beasley, Tatum, and Laurence (1974), Bennett et al. (1970), Gentry (1971), Gentry and O'Neill (1974), and Halls, Bellum, and Lewis (1969). The industry traditionally employed the Hoskold formula, Hoskold (1877), or a variant of it at this level.

11. Ibid.

12. For an analysis of the financial considerations see Lindley et al. (1976), Frohling and McGeorge (1975), and Frohling (1970).

13. Brown (1970) describes the nature of such risks.

14. The economics underlying this model are developed in Conrad (1978a) and Conrad and Hool (1979a).

15. See Roff and Franklin (1964) and MacKenzie et al. (1974) for alternative descriptions.

16. See Harris (1970) and Bennett et al. (1970) for a detailed description of the use of this type of data.

17. See Conrad and Hool (1979a) for a proof. For a mining engineer's perspective on the grade-selection problem, see Walduck (1976) and Thomas (1976).

18. Cutoff grades are generally more complicated, but this model is a standard rule of thumb; see Lane (1964).

19. See Conrad (1978*b*) and references cited therein.

20. See Thomas (1973) for a discussion.

21. U. Peterson has noted this to the authors.

22. See Cigno and Hool (1980) for a discussion of this.

3 Effects of Mineral Taxation

In this chapter we investigate the effects of the various types of taxes commonly employed in the United States.[1] The analysis begins with a consideration of the effects of each individual tax on the firm's behavior. Numerical examples are then developed to demonstrate how the impact of a given tax may differ across mines. Finally, the taxes are considered in combination to exhibit the possibility of reinforcing or offsetting effects. This integration is essential in view of the fact that most states do impose a variety of taxes on the mining industry.

As described in the previous chapter, the overall planning problem for a mining operation is segmented in a hierarchical fashion. However, the objective of each segment is the same: to maximize the present value of extracting and processing ore. Since the major concern of tax policy has been the induced extraction distortions, it is appropriate to begin by analyzing the extraction decision.

Extraction

The following notation will be used to facilitate the analysis. In the absence of taxes, discounted profit in any period is given by

$$\Pi_t = \frac{P_t \, \alpha_t \, X_t - C_t(X_t)}{(1 + r)^t}$$

where

Π_t = discounted profit in period t

P_t = price of output in period t

X_t = ore extracted in period t

$C_t(X_t)$ = total cost of extraction and processing in period t (a function of ore and not of output)

α_t = average grade of ore extracted in period t, measured relative to the standard quality of metal output (concentrate)

r = discount rate

Total output sold is defined as $\alpha_t X_t$. This formulation allows for the effect of grade variation on total production. Note also that total cost is a function of the total tonnage of ore processed, independent of the grade.[2] The cutoff grade is calculated as

$$\alpha^* = \frac{MC_t}{P_t}$$

where α^* = cutoff grade

MC_t = marginal cost of extraction

Ores whose average quality is below the cutoff grade will not be extracted since this would result in a reduction in profit; that is,

$$P_t \bar{\alpha}_t - MC_t < 0 \qquad if \ \bar{\alpha} < \alpha^*$$

We now consider in turn the effects of the various mining taxes.

Severance Taxes

Severance taxes are generally of three types: a fixed payment, in nominal terms, per ton of final output; a fixed payment, in nominal terms, per ton of ore extracted; and a proportion of the sales price or total revenue. Twenty-nine states currently employ some form of severance tax. As noted in chapter 1, this popularity has resulted from the relative ease of administration and collection,[3] and the belief that such taxes can be used to collect rents that accrue from extracting a nonrenewable resource.[4]

After-tax discounted profit for a firm which pays a severance tax imposed per ton of output is

$$\Pi_t = \frac{(P_t - \tau) \alpha_t X_t - C_t (X_t)}{(1 + r)^t}$$

where τ = severance tax in dollars per ton of output. If a fixed payment per ton of ore extracted (that is, at the mouth of the mine) is imposed, after-tax discounted profit is

$$\Pi_t = \frac{P_t \, \alpha_t \, X_t - C_t(X_t) - \gamma X_t}{(1 + r)^t}$$

where γ = severance tax in dollars per ton of ore. Finally, an ad valorem tax on total revenue results in discounted after-tax profit

$$\Pi_t = \frac{(1 - \beta)P_t \, \alpha_t \, X_t - C_t(X_t)}{(1 + r)^t}$$

where $\beta \, (< 1)$ = proportion of revenue collected as tax.

The fixed fee per ton of output is seen to decrease the expected price in each period by the same dollar amount, τ. Since the tax is fixed in nominal terms, the present value of the tax decreases through time; that is,

$$\frac{\tau}{(1 + r)^t} > \frac{\tau}{(1 + r)^{t+1}} > \frac{\tau}{(1 + r)^{t+2}}, \ldots,$$

This means that the operator has an incentive to reallocate extraction from the present to the future. (Some states, such as North Dakota, have attempted to offset this effect by allowing the rate of tax to increase with inflation. This type of tax is discussed below.) For instance, a firm which produces the same tonnage of output over the life of the mine before and after the imposition of the tax will have to pay the same amount of tax in nominal terms regardless of when it is extracted. However, the longer the firm can defer extraction, the lower the tax payments will be in real (present value) terms, and consequently the higher the after-tax discounted profit.[5]

This type of tax may also affect the ordering of grade selection if discounted prices are expected to rise. Recall that chapter 2 showed that the firm will extract the best grade of ore when the expected discounted price is greatest. A severance tax on output effectively reduces the price received by the firm by the same nominal amount in each period. Therefore, if the price is expected to rise, such a tax may alter the period with the highest effective (net-of-tax) discounted price; that is, it may well be the case that

$$\frac{P_t}{(1 + r)^t} > \frac{P_{t+1}}{(1 + r)^{t+1}}$$

but, with the tax,

$$\frac{P_t - \tau}{(1 + r)^t} > \frac{P_{t+1} - \tau}{(1 + r)^{t+1}}$$

This will occur if (and only if)

$$\tau > \frac{(1 + r)P_t - P_{t+1}}{r}$$

A fixed fee per ton of ore extracted will similarly induce a reallo-
cation of ore extraction from the present to the future, but will have
no potential effect on the grade-selection profile. Such a tax increases the cost of extraction and processing by a constant amount,
instead of lowering the price received per ton; that is, it is a tax on
both valuable content and waste. If the raw ore were not processed,
the effects of a $10 per ton tax on output would be identical to a $10
per ton tax imposed at the mouth of the mine. However, most
minerals, even coal, when extracted are used as inputs into a processing plant at or near the mine site where the valuable material is sepa-
rated from waste products. Therefore, the lower the proportion of
raw ore sold after processing, the greater the impact of a tax imposed
at the mouth of the mine. Suppose, for example, that it takes two
tons of raw ore to produce one ton of concentrate. Suppose also that
a state is considering a tax of $10 per ton to be imposed either at the
mouth of the mine or on output of concentrate. If the tax is imposed
on concentrate, then for every ton of raw ore extracted the firm must
pay $5. If the tax is imposed on ore, the firm must pay $10.

The effect of an ad valorem output tax is different again from
those of the other two output taxes. This tax reduces the discounted
price received in each period by the same proportion. It will, therefore, have no effect on the grade-selection decision since the ranking
of discounted prices is unaffected. However, this tax may alter the
extraction profile. A proportional tax on revenue will collect a higher
tax per unit the higher the price. This means that the firm can lower

its tax payments on the same tonnage of output by extracting less ore in periods of higher discounted prices and more ore in periods of lower prices. If discounted prices are expected to fall through time, the firm would reallocate extraction from the present to the future as in the per unit case. If discounted prices are expected to be the same, no distortion occurs, while if discounted prices are expected to rise, the firm will extract more ore early and less in later periods.

Each of these three output taxes creates the further distortion of increasing the cutoff grade; that is,

$$\alpha^* = \frac{MC_t}{P_t - \tau} \quad \text{for the fixed-fee tax on concentrate}$$

$$\alpha^* = \frac{MC_t + \gamma}{P_t} \quad \text{for the fixed-fee tax on ore}$$

$$\alpha^* = \frac{MC_t}{(1 - \beta)P_t} \quad \text{for the ad valorem output tax}$$

An increase in the cutoff grade means that marginal ores will now be left in the ground. The total tonnage of economically recoverable reserves is therefore reduced. This phenomenon is known as "tax-induced high-grading" and has been noted extensively in the natural resources literature.[6]

Since each of these taxes is collected regardless of a mine's profitability, the downside risk associated with mining is increased. Given the risk-averse behavior of most mining operations, an increase in risk will generally shorten the life of the mine because of lower investments and high-grading.

A Note on Variable Severance Taxes

Variable per unit severance taxes have been introduced in some states, examples being North Dakota and Minnesota. The usual justification for such taxes is that, because of inflation, revenues from fixed-rate output taxes do not keep up with the real costs of government services. The tax is therefore linked to some price index in order to compensate for such losses. There are two problems with this procedure. First, there is the choice of the price index with which to adjust the tax rate. If the index is a function of the price, the tax increases regardless of costs. Consequently, firms may have to pay higher taxes even though profits are falling. If the index is not related to price (as in North Dakota and New Mexico), the problem is further compounded because a change in the index is only remotely related to the cost and price structure of any one mine. Such indexation will therefore reinforce the distortionary effects noted above; that is, cutoff grades will rise, extraction will be reallocated, and recovery will be reduced.

Property Taxes

Ad valorem property taxation is the oldest form of taxation used by states and localities for collecting revenue.[7] The tax base is usually defined as the market value of reserves. This market value depends on the quantity of reserves remaining in the deposit and the average grade at the time of assessment. The tax paid is a proportion of this value. After-tax discounted profit is then

$$\Pi_t = \frac{P_t \alpha_t X_t - C_t(X_t) - \sigma F(\bar{\alpha}_t \bar{X}_t)}{(1 + r)^t}$$

$$= \frac{P_t \alpha_t X_t - C_t(X_t) - \sigma F(\bar{\alpha} R - \sum_{j=0}^{t-1} \alpha_j X_j)}{(1 + r)^t}$$

where σ = millage rate

$\bar{\alpha}_t \bar{X}_t$ = output equivalent of ore remaining in the deposit in period t

$\bar{\alpha} R$ = output equivalent of total ore reserves before extraction began

$\sum_{j=0}^{t-1} \alpha_j X_j$ = cumulative output of extraction to date

$F(\cdot)$ = assessment function

(Theoretically, $F(\cdot)$ is the present value of remaining reserves. However, such a value is impossible to estimate, and state and local practices vary. We therefore leave this as a general function.)

Note that the assessed value is directly negatively related to previous extraction. This means that once the ore is extracted and sold the tax base, and therefore tax payments, are permanently reduced. This gives the firm an incentive to increase extraction in early periods and decrease extraction in later periods. In effect, the tax serves as a subsidy to rapid extraction because each ton of ore extracted increases after-tax present value by the tax rate times the present value of the taxes that would have to be paid if the extraction were delayed.

This type of tax may also increase mine recovery by lowering the cutoff grade. If a tax must be paid on ores remaining in the ground, it may be cheaper for the firm to extract and process low-grade ore than to pay the tax indefinitely.[8] In the long run, when capacity is variable, lower investment as a result of taxation may lead to an increase in the cutoff grade.

The effects of the tax are complicated by the inability to accurately assess the property. An ideal property-tax base would be the net present value of the deposit. But such a base is impossible to determine since it depends on future prices and costs as well as the geological structure, none of which is known with certainty. To deal with this problem, states have resorted to arbitrary and sometimes complicated methods to calculate the tax. For instance, the Wyoming tax base is 100 percent of gross proceeds at the mine. This tax is, in effect, an ad valorem output tax and has no relationship to the traditional concept of property taxation. Utah bases assessments on 30 percent of capitalized net income for the previous five years. This tax has the opposite effect to the property tax, because the higher the historical profits, the larger the tax. In addition to the obvious problem of using historical values to project future profitability, there is no clear relationship (and, if anything, a negative one) between historical extraction and the current mineral content of the deposit. This problem is particularly acute in the later years of operation when marginal ores are being extracted. A tax based on historical profits will always overestimate the value of the mine in this case. Finally, some states have replaced property taxes with severance and net-proceeds taxes because of the difficulty of fair assessment.

Proportional Profits Taxes and Depletion
Allowances

It has been well established that a pure proportional profits tax does not affect the behavior of the mine once an investment is made. In addition, it does not affect the downside risks because taxes are paid only when profits are positive. However, the profits taxes employed by the federal government and the various states are not pure profits taxes, because of the various deductions which are available to the corporate sector in general and the mining sector in particular.

The most common deduction available to the mining industry is the depletion allowance. There are two types of allowance currently used by various states: cost and percentage depletion.[9] Cost depletion is a fixed nominal allowance per ton of ore extracted. The amount of the deduction is based on estimates of the costs (or present value) of the mineral property and estimates of the recoverable tonnage of ore. In effect, the allowance attempts to allow a deduction for the reduction in reserves on a unit of production basis. After-tax discounted profit with cost depletion is

$$\Pi_t = \frac{P_t \alpha_t X_t - C_t(X_t) - k\{P_t \alpha_t X_t - C_t(X_t) - d\alpha_t X_t\}}{(1 + r)^t}$$

$$= \frac{(1 - k)([P_t + \frac{kd}{1 - k}) \alpha_t X_t - C_t(X_t)]}{(1 + r)^t}$$

where k = profits tax rate

d = cost depletion allowance (in dollars per ton of output)

As shown, the effect of the depletion allowance is to raise the net-of-tax price received per ton by $kd/(1 - k)$, when profits are positive. It therefore acts as a negative severance tax (an implicit subsidy) per ton of extraction and its effects are accordingly the exact opposite of those of the per unit output taxes. (It should be noted that any form of expensing that is deducted on a unit-of-production basis will have this effect. For example, for federal taxes development expenditures may be deducted in this manner.) Extraction tends to be reallocated from the future to the present, cutoff grades will be lower and recovery higher than would be the case without the allowance.

Percentage depletion allows a fixed proportion of current revenue to be deducted to compensate for exhaustion. After-tax discounted profit in this case is

$$
\Pi_t = \frac{P_t \, \alpha_t \, X_t - C_t(X_t) - k[P_t \, \alpha_t X_t - C_t(X_t) - hP_t \, \alpha_t X_t]}{(1 + r)^t}
$$

$$
= \frac{(1 - k)[(1 + \dfrac{kh}{1 - k}) P_t \, \alpha_t - C_t(X_t)]}{(1 + r)^t}
$$

where h = deduction allowance as a proportion $(0 \leq h \leq 1)$ of revenue.

When profits are positive, this allowance increases the after-tax price of output by a fraction $kh/(1 - k)$ and is, in effect, a negative ad valorem output tax; that is, a subsidy for extraction. (The total deduction is usually limited to some proportion of profits. If that limit is reached consistently, no distortion will occur.) Therefore, the effects of the allowance are the opposite of those discussed with respect to ad valorem output taxes. In addition to these effects, it should be noted that in most cases this allowance is not limited to the total costs of the mine. This means that it is entirely possible for the sum of the deductions taken through time to exceed the costs of acquisition and development. Most of the debate over percentage depletion has been in the context of oil and gas. However, the same analysis applies for nonfuels.[10]

Progressive Profits Taxes

Renewed interest in progressive profits taxes has occurred for at least three reasons. First, governments have sought a means to increase their share of profits in good years (perceived windfalls) without imposing excessive taxation in lean years. Second, governments have been concerned about equity issues, particularly with respect to small marginal operations. In effect, the ability-to-pay principle has been used to argue that large, more efficient firms should be made to bear a larger proportion of the tax burden.[11] Third, there has been a search for alternative taxes which do not have the distortionary effects or large administrative costs of other types of tax which have been in use (particularly ad valorem property taxes). The recent Wisconsin net-proceeds tax is a case in point.

After-tax discounted profit in any period with a progressive profits tax is defined as

$$\Pi_t = \frac{P_t\, \alpha_t\, X_t - C_t(X_t) - T(P_t\, \alpha_t\, X_t - C_t(X_t))}{(1 + r)^t}$$

where $T(\cdot)$ = profits taxes paid in period t, a function of the level of profits, with $T' > 0$, $T'' > 0$. This form of the profits-tax function shows that taxes paid by the firm increase with profits at an increasing rate. In effect, the firm chooses its own marginal tax rate in each period.

The effects of this type of tax depend on the firm's expectations regarding the time paths of discounted prices and costs. In general, the tax will induce the firm to shift output (and therefore profits) from periods with higher tax rates to those with lower rates. There are several ways of accomplishing such a transfer. Extraction can be lowered in periods with high expected profits and increased in other periods; the average grade of ore extracted can be increased or decreased; or the firm can change both extraction and grade to reduce taxable profits in some periods.[12]

Regardless of the method employed, it is important to emphasize the distortionary incentives that are present in such a system. It is unlike a proportional profits tax in that the firm, by choosing the quantity-quality mix, can change its marginal tax rate in each period and thus the total taxes that it will pay. The exact magnitude of the distortion will vary according to the specific design of the tax, but this incentive will always exist.[13]

Some Numerical Illustrations

To illustrate how geological and economic factors combine to determine the tax response of a particular mine, two hypothetical mineral deposits are considered.[14] The essential characteristics of each deposit are detailed in table 3-1. Mine I is composed of a homogeneous ore body, while mine II has three distinct grades with tonnages varying by grade. The marginal costs per ton of ore extracted are the same for each mine, but the fixed costs (and thus the average costs) of extraction and processing are different. Such differences may be due to different techniques or different geological structures. Two alternative price profiles are also noted in table 3-1. Calculations will be made for the two different price profiles to show how alternative economic conditions can affect the outcome qualitatively as well as quantitatively.

The optimal extraction profiles for mines I and II under both price profiles (and in the absence of any taxes) are reported in tables 3-2 and 3-3, respectively. As shown, the optimal life of mine I is three years with either profile. Extraction steadily declines when prices are constant in nominal terms, while extraction is highest in the second year when nominal prices increase in the second and third years. The optimal life of mine II is only two years with either price sequence and none of the 1 percent ore is ever extracted or processed (that is, it is below the cutoff grade). The best grade of ore is extracted and exhausted in the first year.[15] The 2-percent ore is extracted in both years and exhausted. Note that, when the nominal price is increased, extraction of the 2-percent ore is reduced in the

Table 3-1
Hypothetical Mines and Price Profiles

A. *Description of mines to be used in calculations*

Mine I

Grade: 2%	Minimum of average cost curve:
Tonnage: 1,000 tons ore (20 tons output)	$51.38 @ 256.90 tons ore
Total cost $= \$6,600 + .1X_t^2$	Marginal cost: $.2X_t$

Mine II

Grade tonnage	Minimum of average cost curve: $90
4% 450 tons ore (18 tons output)	@ 450 tons ore
2% 550 tons ore (11 tons output)	Marginal cost: $.2X_t$
1% 1,000 tons ore (10 tons output)	

Total cost $= \$20,250 + .1X_t^2$

B. *Alternative price profiles to be used in calculations*
1. $P_t = \$5,000$/ton output every year
2. $P_2 = P_3 = \$5,250$/ton output
   $P_t = \$5,000$/ton output all other years

Table 3–2
Optimal Extraction Profiles for Mine I under Various Price Profiles

Optimal extraction profile when current price is $5,000/ton each year

Year	Extraction (ore)	Output (metal)	Undiscounted Profit	Discounted Profit
1	348.94	6.98	16,118.09	16,118.09
2	333.84	6.68	15,639.09	14,217.35
3	317.22	6.34	15,059.15	12,445.58
Totals	1,000.00	20.00	46,816.93	42,781.02

Optimal extraction profile when current prices are: $P_1 = 5,000, P_2 = P_3 = \$5,250$

Year	Extraction (ore)	Output (metal)	Undiscounted Profit	Discounted Profit
1	333.84	6.68	15,639.09	15,639.09
2	342.22	6.84	17,621.65	16,019.68
3	323.94	6.48	16,919.99	13,983.46
Totals	1,000.00	20.00	50,180.93	45,642.23

Table 3-3
Optimal Extraction Profiles for Mine II under Various Price Profiles

Year	Extraction (ore)	Average Grade	Output (metal)	Undiscounted Profit	Discounted Profit
Optimal extraction profile when current price is $5,000/ton in each period					
1	500.00	.038	19.00	49,750.00	49,750.00
2	500.00	.020	10.00	4,750.00	4,318.18
Totals	1,000.00		29.00	54,500.00	54,068.18

Year	Extraction (ore)	Average Grade	Output (metal)	Undiscounted Profit	Discounted Profit
Optimal extraction profile when current prices are: $P_1 = \$5,000$, $P_2 = P_3 = \$5,250$					
1	488.10	.0384	18.752	49,735.84	49,735.84
2	511.90	.020	10.248	7,295.34	6,632.13
Totals	1,000.00		29.000	57,031.18	56,367.97

first year and increased in the second. This increases the average grade of ore extracted in the first year but decreases total output in that year.

The taxes considered are three forms of output taxes: a 10-percent ad valorem tax on output; a tax of $500 per ton of concentrate produced; and a tax of $10 per ton of ore extracted. These numbers are used for illustration only. However, the distortions will occur regardless of the size of the tax. The effects of each tax on the extraction profile of mine I are reported in table 3-4. When nominal prices are $5,000 in each year the taxes have identical effects, as shown in part A of the table. There are two reasons for this. One is that, when prices are $5,000 per ton, a 10-percent ad valorem tax collects $500 per ton from sales. Therefore, when prices do not change through time a per unit tax is equivalent to an ad valorem tax.

The second reason for equivalence in this case is the uniform-grade distribution. Since the ore is 2 percent throughout the deposit, it takes fifty tons of ore to produce one ton of concentrate. Thus a $10 tax per ton of ore is equivalent to a $500 tax per ton of output in this case. This remains true when prices vary through time. Thus part C of table 3-4 shows again that the $500 per ton output tax and the $10 per ton extraction tax are equivalent. However, a comparison of the two per unit taxes with the ad valorem tax (part B of table 3-4) shows that the output and ad valorem taxes are no longer equivalent. When prices increase the ad valorem tax collects more revenue per ton than either of the per unit taxes, and the response of the firm is correspondingly different, as shown.

Finally, each tax induces the firm to reduce extraction in the early periods and increase it in the later periods. Note that the life of the mine is increased by one year in the constant-price case, while more ore is extracted in year three when prices vary. This conservation effect is the best-known property of such taxes. However, it is only valid when grades do not vary.

The effects of the three output taxes are quite different again when the grade varies within the deposit. The results for mine II are reported in table 3-5. Part A of the table shows the optimal extraction profile with the 10 percent ad valorem tax and the $500 per ton output tax in the constant-price case, while part B reports the profile which results from imposing the $10 per ton tax under the same price

Table 3–4
Effects of Output Taxes on Extraction Profiles of Mine I

A. Optimal extraction profile for mine I with constant prices (equivalent tax case)

Year	Extraction (ore)	Output (metal)	Undiscounted Profit	Discounted Profit	Discounted Tax Revenues
1	277.62	5.55	10,678.51	10,678.51	2,775.00
2	260.39	5.21	10,054.80	9,140.73	2,368.18
3	291.42	4.83	9,299.44	7,685.49	1,995.87
4	220.57	4.41	8,386.19	6,300.67	1,656.65
Totals	1,000.00	20.00	38,418.94	33,805.40	8,795.70

B. Optimal extraction profile for mine I when $P_1 = \$5,000$ and $P_2 = P_3 = \$5,250$ for 10% ad valorem tax

Year	Extraction (ore)	Output (metal)	Undiscounted Profit	Discounted Profit	Discounted Tax Revenues
1	330.66	6.61	12,225.80	12,225.80	3,305.00
2	341.23	6.82	14,002.44	12,729.49	3,255.00
3	328.11	6.57	13,640.78	11,273.37	2,850.62
Totals	1,000.00	20.00	39,869.02	36,228.66	9,410.62

C. Optimal extraction profile for mine I when $P_1 = \$5,000$ and $P_2 = P_3 = \$5,250$ for output tax = \$500/ton and extraction tax = \$10/ton

Year	Extraction (ore)	Output (metal)	Undiscounted Profit	Discounted Profit	Discounted Tax Revenues
1	329.15	6.58	12,189.53	12,189.53	3,290.00
2	342.07	6.84	14,195.47	12,904.97	3,109.09
3	328.78	6.58	13,824.47	11,425.18	2,719.01
Totals	1,000.00	20.00	40,209.47	36,519.68	9,118.10

Note: Taxes are (1) 10% value, (2) $500 ton/output, and (3) $10 tor/ore.

Table 3–5
Effects of Output Taxes on Extraction Profiles of Mine II

A. *10% output tax: and $500/ton output tax*

Year	Extraction (ore)	Average Grade	Output (metal)	Undiscounted Profit	Discounted Profit	Discounted Tax
1	450	.04	18.00	40,500	40,500	9,000.00
2	450	.02	9.00	0	0	4,090.91
Totals	900		27.00	40,500	40,500	13,090.91

B. *$10/ton ore: constant prices*

Year	Extraction (ore)	Average Grade	Output (metal)	Undiscounted Profit	Discounted Profit	Discounted Tax
1	450	.04	18.00	45,000.00	45,000.00	4,500.00
2	450	.02	9.00	0	0	4,090.91
Totals	900		27.00	45,000.00	45,000.00	8,590.91

C. *10% output tax: $P_1 = \$5,000$ and $P_2 = P_3 = \$5,250$*

Year	Extraction (ore)	Average Grade	Output (metal)	Undiscounted Profit	Discounted Profit	Discounted Tax
1	450.00	.04	18.00	40,500.00	40,500.00	9,000.00
2	472.50	.02	9.45	2,075.62	1,886.93	4,510.23
Totals	922.50		27.45	42,575.62	42,386.93	13,510.23

D. $500/ton output: $P_1 = $5,000$ and $P_2 = P_3 = $5,250$

Year	Extraction (ore)	Average Grade	Output (metal)	Undiscounted Profit	Discounted Profit	Discounted Tax
1	450.00	.04	18.00	40,500.00	40,500.00	9,000.00
2	475.00	.02	9.50	2,312.50	2,102.27	4,318.18
Totals	925.00		27.50	42,812.50	42,602.27	13,318.18

E. $10/ton ore: $P_1 = $5,000$ and $P_2 = P_3 = $5,250$

Year	Extraction (ore)	Average Grade	Output (metal)	Undiscounted Profit	Discounted Profit	Discounted Tax
1	450.00	.04	18.00	45,000.00	45,000.00	500.00
2	475.00	.02	9.50	2,312.50	2,102.27	4,318.18
Totals	925.00		27.50	47,232.50	47,102.27	4,818.18

assumption. Note that the extraction profiles are identical in parts A and B. Note also that extraction is reduced in both periods. The best grade of ore is extracted and exhausted in the first year while only 450 tons of 2-percent ore are extracted in the second year. This occurs because the net-of-tax price received per ton of 2-percent ore is exactly equal to the cost of extraction and processing at the minimum of the average cost curve; that is, 2 percent is now the cutoff grade. This is confirmed by the fact that after-tax profit is equal to zero in the second year when only the 2-percent ore is being extracted. The mine operator will be indifferent between extracting 450 tons and extracting nothing at all. Any other tonnage would produce a loss. If the tax had been higher, none of the 2-percent ore would have been extracted.

Although the extraction profiles are identical under either tax, the tax revenues to the government are lower when there is a $10 per ton tax on extraction. This is because of the grade variation. Note that $9,000 in tax revenue is generated in the first year when the tax is imposed on output, while only $4,500 is taken when the tax is imposed on extraction. When ore is 4 percent, only twenty-five tons of ore are necessary to produce one ton of concentrate. Therefore, an equivalent output tax on 4-percent ore would be $250. However, the tax revenues are equivalent when the 2-percent ore is extracted. Again it takes fifty tons of 2-percent ore to produce one ton of concentrate, so that a $10 per ton tax on ore is equivalent to a $500 per ton tax on output. This demonstrates that a per unit tax on extraction imposes a smaller burden on high-grade than on low-grade ores. The lower the grade, the smaller the valuable content and therefore the larger the tax in terms of its value.

Parts C, D, and E of table 3-5 show how the effects of the taxes vary when nominal prices are not constant through time. As shown, the ad valorem tax has a greater effect on the extraction profile than do the two per unit taxes. Again, when prices change, the dollar value of the tax paid changes, altering the effect on extraction. The extraction profiles for the two types of output taxes are identical (the distortionary effects of the taxes are the same), but the tax revenues differ for the reasons cited above.

These examples, while hypothetical, highlight the important fact that economic and geological factors in conjunction determine the consequences of any tax or set of taxes. In actual situations, the effects of a tax across mines could range from little or no effect to a complete closing of an operation, depending on the geological structure unique to each deposit and the general economic conditions.

Integration of Taxes

Several states and localities impose more than one tax on the mining industry. These taxes are used for a variety of purposes, and each tax has its own set of distortionary effects. It is thus of interest to determine how these taxes in combination affect the firm's behavior. Consider first the effects of a profits tax with cost depletion combined with a per unit output tax. If the output tax is not deductible from income for profits-tax purposes, discounted after-tax profit in any period is

$$\Pi_t = \frac{(1 - k)[(P_t + \frac{kd}{1 - k}) \alpha_t X_t - C_t(X_t)] - \tau \alpha_t X_t}{(1 + r)^t}$$

$$= \frac{(1 - k)[(P_t + \frac{kd - \tau}{1 - k}) \alpha_t X_t - C_t(X_t)]}{(1 + r)^t}$$

The depletion allowance increases the net-of-tax price while the per unit tax decreases the net-of-tax price. This means that, used in combination, the allowance and the output tax tend to offset one another. If $kd = \tau$, then they exactly offset each other and no distortionary effect occurs, since the profits tax is neutral in the short run. If a per unit tax were used with percentage depletion, the distortion effects would again be offsetting.

If the severance tax is allowed as a deduction for profits-tax purposes, then for a complete offset the unit tax would have to be equal to the allowance. For example, if the profits-tax rate were 50 percent and cost depletion $10 per ton, a $5 per ton output tax which is not deductible and a $10 per ton output tax which is deductible would each result in no distortionary effect in the short run. The analogous offsetting would result with percentage depletion and an ad valorem severance tax with the rates suitably chosen.

A combination of taxes often employed is an output tax and an ad valorem property tax. When the output tax is ad valorem, after-tax discounted profit is

$$\Pi_t = \frac{P_t \alpha_t X_t - C_t(X_t) - \sigma F(\bar{\alpha} R - \sum_{j=0}^{t-1} \alpha_j X_j) - \beta P_t \alpha_t X_t}{(1 + r)^t}$$

If prices are not expected to rise at a rate greater than the rate of interest (as is usually assumed), the output tax tends to reduce extraction in early periods, which will tend to offset the tendency of the property tax to increase early extraction. There is no simple formula for a combination which will produce no distortionary effects, because the base of the property tax is constantly changing. A complete offset in each period would require changing rates. Note that if a per unit output tax were employed in combination with a property tax, there would be an offsetting tendency regardless of the expected rate of increase in prices.

Another combination of taxes used is a profits tax with a form of depletion allowance and the ad valorem property tax. In this case, both the depletion allowance and the property tax tend to increase extraction in early periods. They complement each other and thus induce faster exhaustion than either tax individually.

In summary, the effects on extraction of taxes used in combination can be offsetting or reinforcing, depending on the nature of their respective distortions. This does not imply, however, that a government can select a set of taxes with offsetting effects and raise an arbitrary amount of revenue. In order for there to be no distortion in the long run, after-tax profits to the firm must remain sufficiently large to justify future extraction. If a sufficiently high level of taxation is imposed the firm may prefer to cease operations entirely.

A Note on Taxation and the Concentrating Decision

The previous analysis has implicitly assumed that the quality of final output is exogenously determined.[16] In practice, the level of processing is part of the decision structure and can be affected by taxation. The concentration (or processing) stage of a mining operation involves the separation of valuable material from waste. In general, a trade-off must be made between increasing the quality of concentrate produced and the total recovery of metal from the raw ore. Efficient operations can usually increase the quality of concentrate only by decreasing the amount of metal recovered.

Along with the geological and geochemical properties of the deposit, economic variables play an important role in determining the trade-off between quantity and quality of output. The price received is a function of the quality, while costs change with intensity of processing.[17] A per unit output tax reduces the quality of concentrate while increasing quantity, whereas an ad valorem output tax has the opposite effect. Finally, neither pure profits taxes nor per unit extraction taxes distort the concentrating decision.

These results are important because one of the goals of natural resource policy has been to ensure maximum recovery from the deposit. Maximum recovery means not only extracting the greatest possible tonnage of ore from the ground. If valuable material is unnecessarily discarded at the processing level, maximum recovery is not being achieved. Care must therefore be taken in designing tax policy so that losses are minimized at both the extraction and concentrating levels.

Effects of Taxation on Investment and Related Issues

A firm considering an investment in a mining operation treats any form of taxation as a cost of doing business in a particular jurisdiction. The taxes paid reduce the present value of cash flow and thus make the investment less attractive. This means that, in addition to the possible effects of a tax (or combination of taxes) on the extraction profile, the tax will tend to decrease investment.

To understand this process, consider the following expression for the net present value of a mineral investment

$$NPV = -I + \sum_{t=0}^{T} \frac{\Pi_t^g - T_t}{(1 + r)^t}$$

where NPV = net present value

I = size of initial investment

Π_t^g = gross-of-tax cash flow in period t

T_t = tax revenue collected in period t

From this simple formulation of the net present value it is clear that an increase in tax payments reduces the present value and makes the investment less attractive. This is true of profits taxes as well as other taxes discussed earlier. It is true that a pure profits tax will not affect the extraction profile, but that is not the only issue when investments are being determined. For example, there is little doubt that for the same investment the firm would prefer an output tax which collects $1,000 in present value terms to a nondistortionary profits tax that collects $2,000 in present value terms.

Any tax will therefore tend to decrease the size of investment in mining activities. This reduction increases the unit cost of extraction, decreases the tonnage of economically recoverable reserves, and generally shortens the expected life of the mine. The tendency to discourage extraction is reinforced by uncertainty. Taxes reduce the expected profitability and thus reduce the investment for a given level of risk. Also the same level of taxation will have a greater deterrent effect the more risky the investment, other things equal. This means that deposits which have low returns but are relatively safe investments may be preferred to deposits which have very high potential returns but have correspondingly high risks.

Expensing Provisions

Mining firms are able to take advantage of special provisions regarding exploration and development, in addition to depletion and depreciation, in calculating profits taxes. In general, these deductions do not change the total taxes paid over the life of the mine. Rather, they change the timing of tax payments from early to later years, thereby decreasing the present value of the tax payments. Some of the justifications for these privileges are (1) mining is more risky than other investments; (2) development is mine-specific with no salvage value; and (3) firms need to finance development with debt in early years and therefore need a sufficient cash flow to repay principal and interest.

There is no doubt that the benefits from exploration are uncertain. However, from an economic perspective, exploration is an investment made by a firm to obtain a stream of future benefits. Like all economic agents, mining firms choose a strategy that will offset these risks. Firms can reduce risks by (1) forming exploration companies that legally separate the risks from other sources of income; (2) pooling risks, by buying interests in property owned by another firm and selling partial interests in their own; and (3) exploring a number of deposits in a variety of areas at one time, to spread the risk within the company. Given this ability to adjust for uncertainty, excessive governmental tax inducements will either transfer resources from other sectors of the economy to mining, or extend exploration to submarginal prospects.

Development, like exploration, is a capital expenditure made to capture a future stream of benefits. As such, good accounting and economic practice dictate that such expenditures be amortized or depreciated to offset these costs as benefits are received.[18]

One of the objectives of granting liberal allowances for exploration and development is to attract mining investments into a particular area. However, in a market economy the benefits may not affect the operations at all, but transfer income to other parties. For instance, a mine may sell the product at a lower price in order to ensure a market. In this case, the firm will not change its behavior, and the tax benefits will be exported with price reductions to other states.[19] The firm may also increase the price it is willing to pay for the right to extract the resource. Again a tax incentive may not increase investment at all but instead be shifted to somewhere or something else. In sum, the effect of these incentives is diluted once the interactions in a market economy are taken into account.

Taxing Rents

In most countries legal title to subsurface rights has been vested in the state. The United States is exceptional in this respect, since title to subsurface minerals is usually recognized as part of the ownership of surface rights. This aspect of the system has a number of implications for tax policy.

First, the original owner of the property collects a payment from the extractive firm for the right to explore and develop his property.

That is, the resource owner collects a "rent" because of his owner-ship. Competition would ensure that this payment would vary across mineral properties according to quality, quantity, and ease of extraction. Because of uncertainty, most of these payments are not lump-sum. Instead, a complicated system of lease bonuses and percentage royalties have developed, to enable some risk sharing between the owner and producer. However, the essential point is that natural-resource rents accrue at least in part to the initial owner of the mineral rights.

Some analysts have argued that states and localities should collect the rents from resource extraction.[20] Based on the above discussion, taxing rents implies that the incidence of the tax should be borne by the original owner and not by the extractive firm. For instance, if the property were sold outright, the state could tax the owner's capital gain from the sale. The firm's behavior would be unaffected, the supply of the resource would be unaffected, and the state would be collecting the rent.

From the analyses of those who would supposedly tax resource rents, this type of taxation is hardly what they had in mind. A true tax on natural-resource rent would be borne by the state's residents (if they own the land) and not by large firms. The tax could not (and should not) be exported or shifted to purchasers in other states. Therefore, a tax on pure natural-resource rent would involve only a transfer of income from the owners of the property (usually residents of the state or the state itself) to the government.

Part of the confusion over taxing rent is the tendency to correlate large profits with rent. It should be emphasized that large profits may bear little relationship to the rate of return on invested capital. Capital is allocated in a competitive economy until the rate of return from investing in various activities is the same. Modern mining is a highly capital-intensive process. This fact, combined with the long lead time for positive returns, suggests the need for sizeable profits in some years. Thus industry spokesmen have some justification for claiming that a comparison of annual profits, without calculating the net-of-tax rate of return, is misleading and unfair.[21] Given also the payments to the owners for the right to extract ore, any association of profitability (even of so-called bonanzas) with some notion of rent is tenuous at best.

Another consequence of the emphasis on profitability when discussing the taxation of mining firms has been the neglect of the fact that taxes paid by the firm are only part of the tax revenue generated by mining activity. Most states impose income taxes on individuals and firms or collect payments from state-owned lands. Capital gains from the sale of mineral properties and royalties payments which accrue to resource owners are part of the income-tax base. Therefore, states are already imposing a form of taxation on the owners of the resource. Little analysis has been done on the effects of this aspect of taxation on the trade of mineral rights or the behavior of owners. Also no estimates are available for the amount of tax collected in this manner. It is clear, however, that the sum of the direct taxes paid by mining firms will underestimate the tax benefits to the state.

In summary, rent is a payment above the minimum amount necessary to bring forth a given supply of a resource. Since the property rights for minerals are usually vested in private hands, it is the owner who collects this payment without contributing to product. A tax designed to collect natural-resource rent must then, by definition, be borne entirely by the property owner, and would have no effect on the profits of the mining firm.

While simple in concept, the ability of the government to collect rents by taxation is complicated in practice. Because of uncertainty, mining firms will typically collect some of the economic rent that would otherwise go to landowners. However, policymakers must be aware of the implications of and distinctions between a tax on rent and a tax on the return to capital. A rent tax will have no effect on the firm's behavior (because the owners of the firm do not bear the burden of the tax), while a tax on the income to capital will reduce investment.

Summary

The preceding analysis examined in detail the response of the mining firm to the major forms of taxation, at each stage of the mining cycle. If price-taking, profit-maximizing behavior is assumed, only a pure profits tax is nondistortionary in the short run. Further, any tax will, ceteris paribus, discourage exploration, development, and investment by reducing the rate of return to capital. A subsidy will have the reverse effect. It was also shown that taxes applied in conjunction may have offsetting effects; for example, an ad valorem property tax and an ad valorem output tax can offset each other. This does not imply, however, that the firm's long-range strategy will be unaffected. Increases in the level of taxation increase the cost of doing business, regardless of the effect on extraction. Thus as noted above, some decrease in investment will result.

Some writers have claimed that price-taking behavior is inappropriate for the analysis of tax policy at the state level.[22] They argue that a state with large reserves can increase the level of taxation without affecting the industry in that state, since the tax can be exported to consumers in other states. They offer such mechanisms as hold-free clauses in long-term contracts as supporting evidence. This claim may well be justified, but only as a short-term phenomenon. A national firm will always try to shift the tax burden to someone else. Its ability to do so will depend on the relevant elasticities of demand for final goods and inputs. But as long as a state does not hold a complete monopoly of the resource, some response by the firm is inevitable. Contracts do expire and buyers will always seek the least expensive source.[23] States must therefore weigh short-run revenue gains against long-term losses from excessive taxation. In the short run behavior may not change, but in the longer run development and investment will fall.

Another implication of this analysis is the importance of separating the issues of taxing rent and taxing the return to capital. It is true that large, rich deposits collect more rent, but in a market economy it is the owner of the land who collects at least part of the expected value of the rent. So a tax on rent should be borne accordingly by landowners and not capital. Such a tax would have no effect on the firm's level of profitability unless it obtained the mineral rights for free. While rents are easy to define, they are impossible to measure. Invariably, some rent is collected by firms as well as consumers of the product. However, an understanding of how rents are generated and their relationship to profitability, if any, is essential for designing tax policy.

Notes

1. The model used to evaluate these taxes is developed in Conrad (1978*a*) and in Conrad and Hool (1979*a*; 1979*b*). The formal proofs are found in these texts; only the results are reported here.

2. See Conrad and Hool (1979*a*).

3. Lockner (1965) and Stinson (1978) noted that the development of severance taxes resulted from the cost of fairly administering the ad valorem property tax.

4. See Boyle (1977).

5. This effect has been established by Hotelling (1931) and Lockner (1965).

6. References to this effect are found in Gillis et al. (1978) and Conrad and Hool (1979*b*).

7. See Warren (1944) for discussion.

8. See Conrad (1978*b*) for a proof. This is contrary to the short-run high-grading effect claimed by McGeorge (1970).

9. See Peterson (1976) for a discussion.

10. The writings of Harberger (1974), Steiner (1959), Davidson (1970), and McDonald (1967) all pertain to some aspect of this debate.

11. The validity of this point is discussed below.

12. See Conrad and Hool (1979*b*) for details.

13. Ibid.

14. These examples are found in detail in Conrad (1979*a*).

15. This is consistent with the discussion in chapter 2.

16. For a formal analysis of this problem see Conrad (1979*b*).

17. See Conrad (1979*b*) for discussion.

18. See Convery and Conrad (1979) for further discussions.

19. The long-run implication of this is discussed in chapter 4.

20. See Boyle (1977).

21. See Hansen (1977).

22. See Long (1976) and Sorenson and Greenfield (1977).

23. The same shifting can occur with subsidies as well. Davidson (1970) argued that depletion allowances mainly increase the rent payments to landowners, with no effect on the firm's behavior.

4

Policy Implications

In the previous chapter we examined the potential effects of various taxes on the decisions of mining operations. It was shown that taxes will typically alter the pattern of quality and quantity of extraction but, unless excessive, will not force an operation to close completely. Rather, investment and total mineral yield will tend to be smaller and the life of the mine shorter as a result of taxation. In addition to these losses, the citizens of the taxing jurisdiction must bear the costs of administration and enforcement. The choice of tax policy is often complicated by a trade-off between the ease of administration and enforcement and the magnitude of the distortionary effects.

This chapter analyzes some of these trade-offs and offers some policy recommendations for consideration. For illustrative reference, we begin with calculations of the impact of the tax system of a given state. A discussion of the incidence (who bears the burden of the tax) follows. In this context, administrative issues are discussed along with the use of particular taxes to meet specific goals.

Calculating the Net Impact of an Integrated Tax System

The first step in the analysis of tax policy is to obtain a measure of the taxes paid by a single firm. Since states generally use more than one tax, this measure must account for the combined effects of the entire state taxation system. It is also important not to overlook the fact that all state and local taxes are deductible for federal tax purposes.

The best way to illustrate the interactive effects of an integrated tax system is with a numerical calculation for a particular state. The representative calculation presented below is for a uranium mine in the state of New Mexico. New Mexico was chosen because of its variety of taxes imposed on mining.

The taxes imposed on uranium operations in New Mexico are reported in table 4–1. To simplify the calculation, the following assumptions are made.

1. The firm is incorporated in New Mexico and has no multistate operations.
2. The firm owns the deposit.
3. The price of uranium concentrate (yellow cake) is $29.04 per pound.[1]
4. The property-tax assessment is $0.2025 per pound of yellow cake.[2]

The first assumption removes the problem of allocating income to New Mexico, which uses the three-factor formula. The second implies that no royalty payments are made. Royalties are deductible from several other taxes.

Total taxes paid to New Mexico by the mining operation are expressed in general terms as

$$T = t_c(PQ - C(Q) - hPQ - aPQ - rPQ - dPQ - eQ - fQ - T_{-1})$$

$$+ aPQ + rPQ + dPQ + eQ + fQ$$

where P = price of yellow cake

           Q = output of yellow cake

      $C(Q)$ = total cost of operations

           h = federal depletion-allowance rate

           a = severance-tax rate

           r = resource excise-tax rate

           d = conservation-tax rate

           e = continued-care tax rate

           f = property-tax rate

       T_{-1} = New Mexico income taxes paid for the previous year

         t_c = corporate-tax rate

Table 4–1
Taxes Imposed on New Mexico Uranium Mines

Tax	Rate		Remarks
Income	5%		Uses federal tax base;[a] therefore grants depletion allowance of 22%.
Severance	*Value*	*Marginal Rate*	Contracts prior to 1977 with no pass-through clause taxed at 1.25% until 1985. Base is total value.
	0– 5.00	1.0%	
	$ 5.00– 7.50	1.6%	
	$ 7.50–10.00	2.0%	
	$10.00–15.00	3.0%	
	$15.00–20.00	4.0%	
	$20.00–25.00	5.0%	
	$25.00–30.00	7.0%	
	$30.00–35.00	9.0%	
	$40.00–50.00	12.5%	
	$50.00 and over	$3.24	
Resource excise	0.75% of value		Allows deduction for state and federal royalty payments.
Conservation	0.0475% of value		Allows deductions for state, federal, and Indian royalties. Rate is 19/100 times 25% value.
Continued care fund	10¢/lb		Payments made until each mine pays $1 million.
Property[b]	Varies with location		For producing underground mines, rate is total sales less state, federal, and Indian royalties times one-fourth times one-third.

Sources: Written correspondence from state tax administrators; Stinson (1978); Gillis (1979); State tax statutes; Commerce Clearing House: State Tax Guide; Yasnowsky and Graham (1976); and Steering Committee on the Impact of Taxation on Energy Markets, National Academy of Sciences (1979).

Note: This tax table was compiled using information from a variety of sources. The information was cross-checked as far as possible to ensure consistency and to include the most recent tax laws.

[a]Certain adjustments for nonbusiness income.
[b]Capital equipment valued at book value.

Because New Mexico uses federal taxable income as its corporate-tax base, it explicitly allows a deduction for all its output-related taxes, the property tax, and income tax from the last period, in computing its income tax. Thus when profit is positive, the effective payment to the state from each tax (excluding the income tax) is 95¢ per $1.00 actually paid (for example, the firm pays a dollar in property taxes but recoups 5¢ by the deduction). Also because of the federal depletion allowance, 1.1 percent $[(.05)(.22) = .011]$ of total revenue is excluded from the tax base.

Substituting the numerical values for tax rates and prices into the expression for T yields the following formula for the tax paid per additional ton of output.

$$MT = .05(29.04 - MC - .22(29.04) - .035(29.04) - .0075(29.04)$$
$$- .000475(29.04) - .10 - .2025) + .035(29.04)$$
$$+ .0075(29.04) + .000475(29.04) + .10 + .2025$$
$$= \$2.73 - .05(MC)$$

where MT = marginal tax and MC = marginal cost of extraction and processing. This shows that New Mexico will collect $2.73 less 5 percent of marginal cost of extraction and processing, for each pound of concentrate sold.

These taxes are partly offset by their deductibility from the federal tax base, which effectively reduces the taxes paid by the firm to the state by an additional 46 percent. Therefore, the impact of the tax system, net of federal tax deductions, on this firm in New Mexico is $1.47 - .027 (MC)$. The net effect amounts to a tax of 5.06 percent of the value less 2.7 percent of marginal cost, per pound of yellow cake.

This calculation highlights several important issues. First, there is the effect of federal deductibility on the impact of state taxes paid by the firm. By allowing the deductions, the federal government is in effect paying 46 percent of the taxes for the firm through lost tax revenue. The total tax burden of both federal and state taxes is, however, higher than it would have been in the absence of state taxes.

Second, there is the reduction in the state's own tax revenue due to the use of the federal base. Since the state taxes are deductible for federal taxes, they are deductible for state income taxes as well. This revenue loss thus reduces the long-run distortionary effects of the taxes.

Third, the tax system increases the cutoff grade. Suppose that the cost per pound of extracting and processing is $2. In the absence of the tax the cutoff grade would be

$$\alpha^* = \frac{MC}{P} = \frac{2}{29.04} = 6.89\%$$

When all the state and local taxes are incorporated, the cutoff grade becomes

$$\alpha = \frac{MC + e + f}{P[1 - a - r - d + \frac{t}{1 - th}]} - \frac{2 + .3025}{29.04[.9412]} = 8.33\%$$

This amounts to an increase of over 20 percent in the cutoff grade. The last equation shows clearly how the net-of-tax price and costs are affected by the output-related taxes. The ad valorem taxes reduce the net-of-tax price, while the per unit taxes increase the cost per pound processed. Note that the depletion allowance tends to offset the decline in the net-of-tax price. Other things equal, the taxes in combination will reduce current extraction and lower total yield.

The discussion to this point has concentrated on the short-run effects of the tax system. But there will also be long-run effects on investment. When incremental investments are contemplated, the tax will serve to reduce the rate of return. To illustrate, suppose the firm is considering opening a new shaft at the mine. The ore in this new area will produce 1,000 pounds of yellow cake at a cost of $2,000. It will cost the firm $24,000 this year to sink the shaft in order to extract all the ore next year. Assuming a price of $29.04 for yellow cake next year, the internal rate of return in the absence of taxation is

$$r_1 = \frac{PQ - C}{I} - = \frac{27040}{24000} - 1 = .1267, \text{ or } 12.67\%$$

However, when New Mexico taxes are introduced, the rate of return is

$$r_2 = \frac{27040 - 1374}{24000} - 1 = \frac{25666}{24000} - 1 = .0694, \text{ or } 6.94\%$$

The introduction of taxation has reduced the internal rate of return by over 45 percent. Other investment criteria may be used, but the qualitative effect of the taxation will be the same.

While the above example is highly simplified, it serves to illustrate the fact that a tax which is small in relationship to the price (or to operating profit) can have a substantial impact on the rate of return. Several authors made calculations of the impact of such measures as tax payments per ton of reserves, tax payments per ton of output, or the ratio of taxes to current profit.[3] These figures do allow a comparison of tax impacts across states but they are misleading indicators of how a set of taxes can potentially affect the level of investment activity. In the long run it is the rate of return to invested capital which will determine the willingness of capital owners to invest in mineral projects. The appropriate measure for evaluating the long-term impact of a tax system is therefore the net-of-tax rate of return to capital.

Incidence Issues

It has been claimed by several authors that increases in the level of state and local taxation will have little or no effect on the behavior of the mining firm. This argument is based on the presumed ability of the firm to shift the burden of the tax to out-of-state purchasers of the resource, in the form of higher prices.[4] To substantiate this, these authors cited "pass-through" or "save free" clauses in contracts, and increased investment in some states despite increased levels of taxation.[5] The ability of a mining firm or the entire industry within any state to export or shift the incidence of a tax depends on a number of factors.[6] First, the state (or group of states) must have some dominance over the resource; that is, the reserves located in the state must be a substantial proportion of total reserves. However, a dominant position is not sufficient for shifting to occur.

A second factor is the elasticity of demand for the resource. If the demand is inelastic, a substantial portion of the tax can be exported. Minerals are used as inputs into other productive processes and it is generally agreed that in the short run the elasticity for most minerals is quite small. However, the long-run elasticity is substantially larger.[7] This is due to the ability of buyers to sign new contracts with lower-cost producers or to change production methods in order to substitute less expensive alternatives for the taxed resource. This means that through time the ability to export taxes through price increases will become much more difficult.

A third factor which must be considered is the mobility of factors of production. When factors are immobile the part of the tax increase which is not exported may be borne by capital and labor within the state. In the long run both capital and labor can move.

Mines close and the capital and jobs will be exported instead of the tax. Therefore, at least part of the burden will be borne by resource owners and the state through lack of future development.[8]

In summary, states can expect little or no effect from tax increases in the short run. However, in the long run, investment and exploration activity will be lower and jobs will be lost as a result of a tax increase. While this conclusion is simple in concept and inevitable in a market economy, it is difficult to convey to citizens and legislatures because it is largely unobservable. What are observed are profit levels and mineral development when prices increase. The conclusion is then drawn that state taxation has little or no effect on the firm's behavior. As prices rise, mineral development will increase as long as the increase in taxation does not capture all the return. The real issues from a policy perspective are (1) how much more mineral development would occur, (2) how much more would be paid to property owners for minerals rights, and (3) how many more jobs would be created, if the rate of taxation were lower?

A recent study by Genetski and Chin offers some empirical support for our prediction.[9] Using cross-section time-series data, they found that the burden of taxation by state had little effect on state income in the same year. However, when they introduced a three-year time lag to allow for adjustments to tax changes they found that an increase of 1 percent in a state's relative tax burden decreases a state's relative income by 0.5 percent. In the context of minerals, this implies that an increase in taxation in, say, New Mexico relative to other states will encourage capital and labor to move to other states over time, decreasing investment and income in New Mexico and increasing it in other states.

Therefore, state and local tax policy should be formulated in full awareness of the fact that higher taxation will reduce the rate of growth of mineral production, decrease economically recoverable reserves, and cause a relative increase in mineral activity in other states.

Recommendations

A rational mineral tax policy must be based on a well-defined set of criteria. Some goals of mineral tax policy are (1) revenue stability (particularly for schools and local government); (2) ease of administration and enforcement; (3) control of external effects such as pollution and land reclamation; (4) consistency with a state's development goals and the relative importance of mineral development in the achievement of these goals; and (5) neutrality of the tax system.

From the preceding discussion, it is clear that no single tax can simultaneously achieve all these goals. Recent theoretical literature also suggests that more than one type of tax may be preferable to a single tax.[10] For instance, an income tax designed to ensure neutrality between mining and other sectors may be impossible to administer because of its complexity and may not generate adequate revenues in periods of depressed markets. On the other hand, a severance tax per unit of output is a relatively stable source of revenue and is easier to administer, but it induces excessive high-grading and increases downside risks.

In addition, different weights will be assigned to each goal, according to the circumstances particular to each state. Finally, economic conditions change and tax policy should be flexible enough to adjust to new conditions without excessive inducements or disincentives.

States differ with respect to both the type and geological composition of each mineral and their willingness to pay for tax administration. (This is probably one reason why a regional tax policy has never developed.) However, the same basic criteria will apply to each state. Therefore, our recommendations deal with the general structure of the tax system. Particular modifications can then be incorporated by each state according to its specific circumstances.

The Income Tax

A well-designed income (profits) tax should be the basis for any mineral tax policy. From a corporate perspective, the income tax is a levy on the return to invested capital. Therefore, an income tax on corporate income can be designed so that it is neutral with respect to incentives within the corporate sector. The income tax has the further advantage of not increasing the downside risks as much as other taxes, since costs are considered explicitly and taxes are paid only when profits accrue. Third, the tax does not induce high-grading in the short run, so will have relatively little effect in terms of reserve loss. Finally, if there is also an individual income-tax system, the state is assured of collecting part of the natural-resource rents. In effect, the state should not care how the rents are divided between resource owners and extractive firms, because it will collect part of the rent from both sources.

The following provisions should be part of the income-tax package.

1. Only cost depletion should be allowed.
2. Capitalization of exploration expenditures should be required.
3. Development expenses should be subject to usual depreciation rules.
4. All other state and local taxes should be deductible.
5. Taxable income should be defined to include only income accruing within the state (that is, no income allocation rule).

Percentage depletion, as used in the United States, has been shown either to act as an income subsidy for mining firms or to induce excessive investment in mining activity.[11] The previous discussion also showed that the allowance will generally increase early extraction and reduce the life of the mine. Finally, no other industry is permitted to deduct an amount greater than its investments costs. Neutrality therefore dictates the elimination of percentage depletion.

The expensing privileges for exploration and development have been claimed to be justified by the risks inherent in mining and the lead times necessary to recapture the invested capital. However, these risks should be reflected in market pricing. Therefore, good accounting and economic principles imply that such expenses be deducted as income accrues.

The deductibility of other state taxes will allow a partial mitigation of their distortionary effects. In addition, other taxes are generally privilege taxes or a form of use charge for publicly provided goods. They are thus part of the cost of doing business in the state and should be recognized as such.

In theory, a state income tax should only recognize income and costs which accrue in that state. Under the current allocation system used by several states, it is possible that firms which suffer losses in a state must nevertheless pay income taxes there because of profits generated elsewhere. Likewise, firms with high profits in one state may be able to reduce their tax liability there by deducting losses from operations in other states. In addition, McClure (1977) recently showed that the formula currently employed acts to make the income tax a series of taxes on each of the three factors included in the formula (sales, employment, and assets), whose incidence may be regressive. Therefore, to preserve the neutrality and the intended incidence of an income tax, only intrastate income should be taxed.

Some of the above recommendations are at variance with current federal practice. States that currently use federal taxable income as the base will consequently incur a greater administrative burden if they adopt such measures. This is especially true with regard to the last recommendation. States will have to determine what proportion of corporate overhead, research and development, and management fees can be deducted within a state, if any. While some of these decisions will inevitably be arbitrary, the bias with such an allocation will not be nearly as great as under the current system.

The net-proceeds tax, which has become increasingly popular, is similar in effect to an income tax with the characteristics we prescribe. This move toward taxes of the net-proceeds type is an indication that states are willing to bear some additional costs to ensure, as well as increase, their revenues.

Output-Related Taxes

Any output-related tax will discourage investment and induce high-grading. In addition, such taxes increase the downside risks associated with mining since they do not recognize costs. They are, however, extremely popular tax instruments and their use will apparently continue. Given this, our recommendations are intended to serve as guides for minimizing distortions while generating revenue.

First, the tax should be imposed at the earliest possible point, that is, at the mouth of the mine if possible. This will minimize the effect of the tax on other stages, such as concentrating and refining, and will thus partly reduce the incidence on capital and labor.[12]

Second, the tax should be related to the value of the ore. While some high-grading is inevitable there is no justification for imposing a uniform per unit tax on all ore. Such a tax will always create a larger high-grading effect for the same revenue than a tax which is related to value.

Three pieces of information are necessary to administer such a tax: the tonnage, the grade, and the price. The first is easy to obtain, the second may require independent analysis to ensure proper enforcement, and the third may be difficult to obtain. The difficulty in obtaining price information is due to the fact that most metals are not sold in their natural form. Rather, they are inputs into a concentrating process whose output is then marketed. Therefore, an arm's-length price in general does not exist. One solution is to determine first the arm's-length price of concentrate, either from an analysis of the contract provisions or from an independent source, and then allow a deduction for processing charges to arrive at a net price per ton. This procedure will yield the value of ore in concentrate that can serve as the base of the tax. (This is the same method used in determining income from mining which serves as the base for percentage depletion.)

Finally, we recommend that the tax be at a uniform rate. Different rates may be imposed for different minerals or for alternative mining techniques (that is, one rate for underground mines and one for open pits) but the rate should be independent of the price. The usual justification for progressive severance taxes is that firms do nothing to bring about a price increase and thus collect a windfall. This argument has three basic flaws. First, a windfall should be defined only with regard to the rate of return on capital, and not the price. It is entirely possible, given the risks of mining and the price uncertainty, that a price increase may only reflect a minimal rate of return. Second, a price increase may be accompanied by cost increases such that the price increase is associated with lower profits rather than a windfall. Finally, this tax is totally discriminatory. If a state legislature feels it has a right to tax windfalls, it should do so consistently. That is, it should increase the rate of taxation on wheat farmers when wheat prices are high, independent of costs; tax the sale of homes and property more heavily when housing prices are high, regardless of maintenance and upkeep expenses; and so on. Unless it can be proved that such a tax is functionally related to a market failure, its use is discriminatory and should be avoided. As noted above, an output tax may be appropriate for environmental conservation, but the cost of environmental damage bears little relationship to the market price.

If in spite of this such a tax scheme is to be used, it should be related to the price of the specific mineral and should acknowledge price decreases as well as increases. A tax related to the consumer price index (as in New Mexico) or the wholesale price index (as in

North Dakota) has a tenuous relationship to the profitability of any mine. Such indexes are composites of national averages and thus could overestimate or underestimate the effect of the general level of inflation on a particular region or mine. That the tax rate should decline when prices fall is suggested as the natural counterpart to the taxation of supposed windfalls. Such flexibility would reduce somewhat the discriminatory bias of the tax.

Property Taxes

One of the more important goals of state and local tax policy is to provide a stable source of revenue to finance a minimal level of public services such as schools, police, and fire protection. Each of the taxes outlined above is unsatisfactory in this respect because the tax revenue is a function of mineral prices (and costs in the case of income taxes), which have historically been cyclical. Therefore, they cannot be relied on to provide a stable revenue base for the provision of public programs.

Property taxes, on the other hand, do offer such stability. This is due less to the nature of the tax than the way it is administered. Historically, proponents of the property tax have emphasized its revenue stability due to the inelasticity of the base. For example, Maxwell (1965) suggests that local governments resort to the property tax, rather than income or sales taxes, because the immobility of real property precludes a shrinkage of the tax base. This view has been challenged both theoretically and empirically.[13] It has been shown that the supply of real property is inelastic only in the short run. Over time, factors are mobile and an increase in the rate of taxation will encourage movement. Individuals and firms have moved to the suburbs to escape city taxes (while still taking advantage of city services). Local stores followed and shopping centers developed. In addition, high property taxes have been a factor contributing to an exodus from certain regions of the country. Therefore, in the longer run, the tax base is not secure and thus cannot be the explanation for the stability of the revenue generated by the property tax.

The key to the tax's revenue stability lies in its administration. The tax rate is usually determined after the budget for the next fiscal year has been projected. That is, the tax rate is set to collect a specific amount of revenue. Since this is the process typically used, the revenue is stable by definition, quite independent of the base. In this sense, any tax can be a stable source of revenue. Given this, real property may be a preferable base since it is less subject to periodic fluctuations than are other tax bases. But again it must be emphasized that the stability of the base itself is not to be identified with the stability of revenue.

Theoretically, the value of the property is its market value. From an economic perspective the market value of a fixed asset is the present value of income generated by holding or using the asset. Therefore, a property tax based on market value is an income tax which is paid on the future stream of benefits generated by the asset. The difference between the regular income tax and the property tax is that the former is a tax on the realized stream of income, while the latter is a tax on the expected future stream of income. If there were no uncertainty, a property tax could be designed in such a way that the impact and incidence of the tax would be identical to a tax on realized income.[14]

Uncertainty, however, is a complicating factor and the difficulties it creates are most apparent in the mining sector. Unlike other forms of real property the extent of mineralization in a mine is never known, in most cases, until it is extracted and processed. In addition, mines are not subject to frequent sale and the characteristics unique to each mine make comparisons difficult across mines. Finally, mineral-price profiles tend to be more uncertain than those of other types of real property. These uncertainties combine to make property valuation more difficult in this sector.

States and localities have realized this problem and have developed numerous ways to cope with it. Among the tax bases currently in use are (1) a fraction of total revenue (in effect an ad valorem output tax); (2) net proceeds (in effect a tax on current realized income); (3) estimates of the present value; and (4) traditional local assessment. Of these four methods, the estimate of the present value is most likely to reflect the true worth of the deposit. However, the measure is far from perfect and the estimates must be derived with care.

The valuation procedure we recommend is based on the Arizona method, which we feel is both feasible and more accurate than other methods. In this approach, the present value of the operation is calculated on the basis of estimates of reserves and production. Instead of projecting prices and costs, a profit-margin formula is used to convert the extraction profile into an income stream which is then discounted to obtain the present value. The profit-margin formula is based on the average margins for the preceding five years to even out fluctuations.[15]

This method has a number of advantages. First, it is consistent and is based on the same principles as used in the industry itself. Second, the tax base will not discriminate against marginal mines. A marginal mine is, by definition, one with a lower profit margin, and this will be reflected in the calculation. Third, it recognizes costs and thus is a better measure of income from property than are sales. Fourth, it uses a historical average margin which weights the good years with the bad, and is therefore better in this respect than current net proceeds. In addition, it alleviates the burden of projecting future costs and prices. Finally, it is not as distortionary as other taxes. There still exists an incentive to extract more, and higher quality, ore early to keep down the estimated value of future extraction. However, this effect is offset, at least in part, by the use of historical profit margins. If the firm were to maximize the current return, profit margins would be higher and would thus serve to increase the future value.

The method does have one major disadvantage: the cost of administration. In order for the method to be accurate and consistent, much information is necessary and several unknowns must be estimated. Comprehensive field reports are written annually on each operation to keep a current record of technology and capitalization rates. Also statistics are compiled from financial and professional literature and Securities and Exchange Commission reports, in order to assure that assessments are fair. (It is our impression that some problems exist in the Arizona system and that property values in some cases may be subject to arbitrary methods.) It is clear that such a method can only be employed by a state agency with the resources to devote to such a task. However, any state that has an active mining sector should be collecting this type of information at the state level in any event, and so the method could be employed without much additional expense.

Once the base is estimated, the assessment ratio must be determined. In the absence of proved market failure, the assessment ratio should be the same for mining as for other industrial sectors. Some states, Arizona included, employ higher ratios for mining. This distorts the long-run allocation of investment. The onus is on state and local governments to justify such a distortionary incentive which deters development and cuts jobs in the mining sector.

Concluding Remarks

The recommendations outlined above reflect the view that for tax purposes, mineral development should be treated in the same manner as any other form of economic activity. In the absence of proved market failures, this means that the burden of the taxation borne by mining activities should be the same as that borne by other sectors. While it is true, in a sense, that the minerals found within the boundaries of a state or locality are finite, this alone is not a sufficient condition for excessive inducements for or taxes on mineral activity.[16] In the long run, the effect of depletion will be largely beyond the control of any one state. State and local tax laws may distort the current allocation of investment within and between states. However, these relative distortions will not substantially affect the long-run availability, because there are too many alternative sources.

A balanced approach is necessary to ensure that the minerals sector develops along with other sectors, so that the scarce resources can be put to their most productive use. If market failures are present, the state has the obligation to correct them, and tax policy may be an efficient means by which to modify the system. However, the basis of the tax system should be neutrality. Otherwise, a proper evaluation of the effects of other adjustments cannot be made.

Notes

1. Average price calculated by the New Mexico Taxation and Revenue Department (14 February 1979).

2. Tax calculated by the New Mexico Taxation and Revenue Department.

3. For one example, see the New Mexico Taxation and Revenue Report, where the impact of taxes is calculated to be about 6 percent. Income taxes and federal deductions have been ignored in this calculation. For another example see Shelton and Morgan (1977).

4. Griffin and Shelton (1978) adopt this argument, while Long

(1976) uses it as a justification for a regional tax policy on coal. See also Link (1978).

5. Sorenson and Greenfield (1977) claim that a state can tax mining and capture at least part of the comparative advantage of a state's mine with no effect. However, for this to occur the tax must be shifted.

6. For a detailed analysis see McClure (1978).

7. See Conrad (1978a) and the references therein.

8. McClure (1978) argues that in the long run the tax will be borne almost entirely by resource owners, because reserves are the least mobile factor.

9. Cited in the *Wall Street Journal,* 9 September 1979.

10. See Atkinson and Stiglitz (1976).

11. See Harberger (1974) for the classical statement of the distortion introduced by percentage depletion.

12. See Conrad (1979b) for a discussion of downstream effects.

13. See Mieszkowski (1972).

14. See Fiekowsky and Kaufman (1976).

15. This procedure is outlined in the Arizona statutes.

16. Long (1977) supports this view.

Appendix
State Tax Collections

Table A-1
State Tax Collections
(thousands of dollars)

Tax	1971	1972	1973	1974	1975	1976	1977	1978 (prelim)
Alabama								
License	53,380	66,834	76,765	77,191	81,620	90,704	99,704	106,046
Oil companies	432	510	559	613	1,038	1,269	1,700	1,957
Individual income (net)	92,094	118,994	142,231	169,801	189,964	224,597	261,895	317,958
Corporation net income	33,764	32,908	40,939	47,255	58,158	60,307	75,874	83,161
Corporation: general	29,667	29,046	36,622	42,283	52,643	55,425	69,043	74,307
Property	24,020	25,355	27,498	26,447	28,593	31,381	34,095	53,877
General	23,735	25,052	26,961	25,884	27,929	30,670	33,212	52,784
Severance	2,379	5,083	6,522	8,627	10,948	11,570	13,757	17,056
Production privilege	994	1,492	2,060	3,632	5,393	6,158	7,686	10,321
Oil and gas	497	606	824	1,794	2,697	3,076	3,843	5,161
Coal tonnage		2,040	2,635	1,771	1,507	843	672	
Iron ore tonnage	19	8	8	4	1			
Alaska								
Individual income (net)	35,484	39,112	43,363	49,219	86,975	156,254	210,338	145,828
Refunds					NA	−30,889	−41,029	−35,513
Gross collections					NA	177,143	251,430	181,341
Corporation net income	6,050	6,455	6,964	8,241	17,345	31,103	35,772	33,504
Property (special)					6,566	306,429	409,767	173,197
Oil and gas reserves						223,147	270,627	
Oil and gas properties					6,566	83,282	139,140	172,995
Severance	14,491	14,905	14,099	17,515	26,619	27,978	23,758	107,715
Oil and gas production			11,469	14,760	26,542	27,901	23,705	107,600
Oil and gas conservation				3	77	77	53	115

Arizona

Sales and gross receipts	316,435	361,950	419,644	466,695	579,665	531,172	710,534	803,503
Mining	9,582	8,465	10,139	12,585	11,560	11,799	14,540	11,744
Individual income (net)	73,710	94,577	108,631	137,698	157,537	162,869	190,591	222,808
Refunds					NA	−17,873	−24,181	−31,427
Gross collections					NA	180,742	214,772	254,235
Corporation net income	26,987	28,126	37,408	39,356	49,553	46,499	51,788	63,842
Property	65,547	68,537	70,392	47,607	97,158	114,689	129,834	136,182
General	57,360	59,016	59,558	35,870	75,354	93,642	102,781	109,253
Special	8,187	9,521	10,823	11,737	21,804	21,047	27,053	26,929

Arkansas

Individual income	44,243	70,150	39,343	117,022	126,192	147,688	163,781	202,939
Refunds					NA	−21,440	−35,349	−36,023
Gross collections					NA	169,128	199,130	238,962
Corporation net income	26,384	31,568	37,825	45,916	54,469	56,197	67,210	83,528
Property	976	1,078	1,213	1,416	1,625	1,779	1,703	2,080
General	103	91	112	121	126	178	153	160
Severance	4,667	4,973	4,911	6,528	7,264	8,870	10,495	12,391
General	3,679	3,871	3,840	5,396	6,287	7,798	9,248	10,632

California

Individual income (net)	1,266,556	1,838,503	1,886,442	1,803,080	2,456,573	2,957,788	3,620,933	4,532,488
Refunds				−455,367	−588,843	−634,766	−864,336	−819,390
Gross collections				2,258,447	3,045,416	3,592,554	4,485,269	5,451,878
Corporation income	533,121	661,071	866,347	1,046,031	1,252,633	1,284,068	1,641,595	2,076,270
Severance	1,830	2,262	2,079	2,537	2,331	2,334	1,530	30,452
Petroleum and gas	1,231	1,554	1,407	1,661	2,331	2,334	1,530	1,530

Table A–1 continued

Tax	1971	1972	1973	1974	1975	1976	1977	1978 (prelim)
Colorado								
Individual income (net)	143,461	174,269	185,785	250,527	280,498	294,008	338,920	375,341
Refunds						−111,630	−116,119	−155,725
Gross collections						405,638	455,039	531,066
Corporation net income	28,837	36,463	38,993	52,745	58,115	68,504	80,575	86,202
Refunds						−8,849	−7,752	−9,393
Gross collections						77,353	88,327	95,595
Property	2,008	2,328	2,561	3,181	1,697	1,665	2,438	2,607
Severance	567	561	833	1,108	2,361	4,371	2,320	1,838
Oil and gas production (gross income)	430	435	63	846	1,932	3,896	1,762	1,370
Oil and gas conservation	94	89	182	217	333	417	482	404
Coal tonnage	43	37	38	45	46	58	76	64
Georgia								
Individual income (net)	183,353	239,900	284,909	340,040	373,916	413,188	495,639	604,361
Refunds				−73,435	−83,889	−91,888	−101,867	−115,676
Gross collections				413,475	457,805	505,086	597,506	720,037
Corporation net income	81,688	88,928	114,114	132,629	119,353	132,741	170,885	203,823
Property	4,057	4,055	4,590	5,099	6,123	6,612	9,903	7,914
Idaho								
Individual income (net)	39,755	50,191	57,691	72,183	91,244	98,824	112,470	138,050
Refunds				−16,074	−22,742	−27,528	−32,506	−27,998
Gross collections				88,257	113,986	126,352	144,976	166,048
Corporation net income	12,563	12,894	16,024	23,076	28,162	31,752	31,034	33,326

Property	446	289	283	372	325	257	228	235
General	380	242	216	295	252	179	148	92
Special	66	47	67	77	73	78	80	143
Severance: mining privilege	268	152	73	192	481	394	203	273
Illinois								
License						400,327	412,505	425,701
Mines and minerals						583	605	437
Individual income (net)	773,610	843,251	894,697	1,046,575	1,136,918	1,216,557	1,413,368	1,593,695
Refunds				−92,432	−110,095	−108,045	−126,546	−72,230
Gross collections				1,139,107	1,247,013	1,324,602	1,539,914	1,665,926
Corporation net income	154,984	173,912	223,083	226,944	306,828	312,131	384,410	376,098
Indiana								
Individual income (adjusted gross-income tax)	218,467	283,669	284,916	328,071	400,793	405,432	479,259	538,225
Refunds				NA	−72,276	−43,995	40,000	−39,475
Gross collections				NA	473,069	449,387	519,259	577,700
Corporation net income	9,580	10,526	10,084	25,542	77,427	85,487	86,198	192,068
Supplemental net income				15,400	52,094	56,387	66,037	122,872
Adjusted gross-income tax				10,242	25,333	29,100	20,161	69,196
Property	15,592	22,226	24,688	25,111	23,046	24,989	23,568	23,650
General	1,148	1,740	945	300	1,760	1,130	2,097	1,892
Special	14,444	20,486	23,743	24,811	21,286	23,859	21,471	21,758
Severance: petroleum production	239	221	203	315	475	508	580	649
Iowa								
Individual income (net)	115,344	202,158	242,863	320,594	358,899	388,212	447,409	490,210
Refunds				−44,206	−51,462	−61,739	−61,204	−74,855

Table A–1 continued

Tax	1971	1972	1973	1974	1975	1976	1977	1978 (prelim)
Iowa continued								
Gross collections				364,800	410,361	449,953	508,613	565,065
Corporation net income	28,359	37,109	47,288	59,416	60,024	77,831	91,894	108,961
Kansas								
Individual income (net)	82,156	95,345	114,268	147,143	170,044	193,730	209,171	241,224
Refunds				−10,551	−12,520	−11,313	−22,287	−31,416
Gross collections				157,694	182,564	205,043	231,458	272,640
Corporation net income	25,112	33,153	53,821	76,766	85,887	92,848	116,721	128,513
Corporations					77,213	84,605	106,301	116,225
Property	10,459	11,330	11,807	12,503	13,294	13,972	9,976	16,561
General	9,091	9,698	10,028	10,445	11,004	11,190	444	13,104
Severance	664	687	711	704	700	785	816	841
Natural gas	405	432	446	464	458	402	516	547
Oil production	138	120	118	112	106	237	149	149
Oil proration	121	135	147	138	136	146	151	145
Kentucky								
License		51,459	57,241	61,398	67,592	72,041	75,139	83,574
Strip mining permits		808	903	1,295	2,594	1,830	1,892	2,862
Individual income (net)	132,669	156,369	179,216	212,324	249,449	292,546	338,160	389,912
Refunds				−44,652	−52,357	−59,320	−66,844	−76,320
Gross collections				256,976	301,806	351,866	405,004	466,232
Corporation net income	40,093	53,903	69,338	83,364	116,626	134,785	131,254	138,597
Property	26,746	38,477	31,385	33,681	35,364	37,299	43,413	147,366
General	2,285	2,430	2,405	2,688	2,912	3,216	3,448	75,514

Severance	182	5,941	37,385	53,749	99,089	91,496	113,005	128,160
Coal		5,767	37,226	53,495	98,740	91,078	112,597	127,765
Oil production	182	174	159	253	349	418	408	395
Louisiana								
License	70,727	79,422	80,499	92,876	97,494	114,158	122,636	140,725
Drill and renewal permit	517	416	405	331	390	424	533	544
Individual income (net)	81,867	105,354	109,417	99,556	108,870	117,641	133,614	192,276
Refunds						NA	-34,957	-24,314
Gross collections						NA	168,571	216,590
Corporation net income	79,523	51,299	78,781	67,603	78,718	87,741	95,248	186,956
Property: general	29,117	28,254	24,030	163	26	234	147	34
Severance	256,600	244,456	267,712	390,345	548,510	358,495	495,498	476,829
Oil and distillate	132,984	122,527	112,061	186,061	282,483	275,928	269,874	259,614
Gas	116,095	113,958	147,394	192,214	250,940	225,792	210,372	202,031
Sulfur	2,659	2,459	2,242	2,131	2,046	4,062	2,435	2,295
Maine								
Individual income (net)	23,878	28,179	31,308	39,033	44,603	52,190	75,157	103,177
Refunds				-8,749	-11,030	-12,390	-19,781	-19,114
Gross collections				47,782	55,633	64,580	94,938	122,291
Corporation net income		8,558	10,044	13,202	20,181	32,642	35,200	34,307
Property	3,967	5,819	5,993	5,776	10,250	13,496	14,317	18,763
General	3,435	4,885	5,251	5,771	10,238	13,480	14,295	18,741
Maryland								
Individual income (net)	413,976	456,854	515,933	573,728	665,997	790,364	806,740	884,392
Refunds					NA	-181,803	-197,081	-226,872
Gross collections					NA	972,167	1,003,821	1,111,264

Table A–1 continued

Tax	1971	1972	1973	1974	1975	1976	1977	1978 (prelim)
Maryland continued								
Corporation net income	70,260	77,441	80,035	90,065	94,389	109,254	115,297	126,802
General corporation					83,968	98,404	101,514	112,528
Property								
General: loan taxes	33,541	36,273	45,954	50,047	52,419	59,792	72,446	80,437
Special: rolling stock	33,424	36,162	45,647	49,738	52,085	59,486	72,143	80,094
Interest and penalties	118	112	115	109	101	34		55
			192	198	234	272	303	288
Minnesota								
Individual income	370,702	483,215	586,235	701,398	807,108	849,520	956,933	1,074,552
Refunds				−157,641	−169,485	−216,845	−254,257	−311,465
Gross collections				859,030	976,593	1,066,369	1,211,200	1,386,017
Corporation net income	79,969	112,403	170,655	190,326	195,905	196,436	258,095	292,853
Refunds				−13,408	−19,269	−28,100	−27,038	−27,765
General (gross)				187,215	199,004	203,256	263,737	293,646
Property								
General	6,844	2,868	2,122	2,576	2,355	2,182	3,083	3,760
		139	48	55	22	14	22	18
Severance								
Occupation tax	18,388	20,080	19,924	29,394	35,897	58,171	59,718	61,945
Taconite	9,312	7,292	10,022	15,842	20,055	24,321	25,057	10,193
Iron ore			6,358	6,885	10,235	19,218	18,141	6,247
Royalty tax			3,664	8,958	9,820	5,103	6,916	3,946
Taconite	1,647	1,840	2,900	3,392	3,890	3,503	3,793	2,853
Iron ore			2,477	1,953	2,356	2,770	3,039	1,905
Copper, nickel			421	1,437	1,532	731	752	945
			2	2	2	2	2	3
Taconite tonnage and additional tax	7,429	10,948	7,002	10,159	11,952	30,347	30,868	48,889

Missouri								
Individual income (net)	168,932	256,801	315,027	315,481	311,334	333,843	389,594	438,604
Refunds				−44,461	−61,037	−62,741	−72,439	−91,983
Gross collections				309,942	372,371	401,584	462,033	530,587
Corporation net income	27,322	50,012	62,664	54,683	56,405	83,680	105,772	111,953
Corporation income					56,405	67,439	87,160	91,224
Refunds					−5,844	−9,588	−6,930	−7,689
Gross collections					62,249	77,027	94,090	98,913
Property: general	3,252	3,899	3,558	3,802	3,952	4,364	4,492	4,627
Montana								
Individual income (net)	42,381	68,082	77,056	79,029	88,599	97,520	111,862	123,621
Refunds				−11,902			−14,212	−18,317
Gross collections				90,931			126,074	141,938
Corporation net income	9,546	11,523	12,507	15,736	22,079	23,020	24,957	29,239
Property	8,596	7,492	6,694	22,507	10,604	16,813	15,636	16,329
General				22,466	10,546	16,737	15,636	16,249
Severance	5,131	4,474	5,229	9,822	14,685	31,344	45,753	44,667
Coal production	2,933	2,668	2,698	3,315	5,396	22,924	34,470	33,624
Oil production	212	483	694	4,256	6,180	6,564	6,884	6,808
Resource indemnity							2,210	2,246
Metal mines tax	1,977	1,314	1,828	2,240	3,099	1,845	2,178	1,979
Nevada								
Property	11,479	12,621	14,617	16,581	17,586	19,964	22,105	27,051
General				6,576	7,103	8,210	8,287	10,215
Real property			5,250	6,104	6,538	7,572	7,576	9,217
Personal property			682	472	565	638	711	998
Severance: mine proceeds	50	128	104	156	177	148	105	129

Table A–1 continued

Tax	1971	1972	1973	1974	1975	1976	1977	1978 (prelim)
New Jersey								
Individual income (net)	19,570	23,258	25,527	44,035	45,942	101,200	709,653	778,505
Corporation net income	112,312	119,528	170,588	197,591	202,780	228,996	332,775	389,227
Corporation income tax							314,323	379,782
Property: Special: business personal property	50,842	53,442	57,745	64,235	70,742	64,742	80,491	81,176
New Mexico								
Individual income (net)	35,815	44,088	49,501	57,946	56,575	58,191	26,639	45,992
Refunds				−15,564	−17,214	−23,324	−72,116	−55,834
Gross collections				73,510	73,789	81,515	98,775	101,826
Corporation net income	10,119	13,211	15,063	16,610	18,344	23,504	29,486	37,608
Property								
General	15,893	15,030	13,230	13,833	14,466	13,841	16,095	19,851
Special: oil and gas					13,994	13,210	13,489	19,023
production					472	631	2,606	828
Severance	35,815	35,878	36,947	43,963	71,154	87,485	102,783	145,826
Oil and gas, 3¾%				18,390	36,064	46,211	53,975	74,088
Oil and gas privilege				18,752	26,250	31,457	36,715	42,831
Oil and gas ad valorem production				3,246	3,906	4,058	4,419	5,034
Oil and gas conservation				1,030	1,456	2,225	2,787	3,515
Natural gas processors				1,319	1,695	2,199	2,719	3,210
Other: copper, potash, uranium				1,226	1,513	1,335	2,168	17,148

New York

Individual income (net)	2,550,207	2,514,557	3,221,930	3,431,993	3,588,584	3,948,808	4,526,975	4,506,245
Refunds				-713,631	-758,240	-960,611	-576,783	-1,138,573
Gross collection				4,145,624	4,346,824	4,639,419	5,103,758	5,644,818
Corporation net income	572,328	781,010	874,267	874,379	967,401	1,132,756	1,295,001	1,344,610
Corporation franchise	433,825	601,369	693,948	706,174	763,269	877,190	1,048,021	1,080,596

North Dakota

Individual income (net)	16,877	19,506	27,318	45,435	67,649	50,477	55,037	69,171
Refunds					-3,095	-6,143	-6,950	-5,387
Gross collections					70,743	56,620	61,987	74,558
Corporation net income	7,723	8,872	10,089	14,526	19,964	19,572	21,800	20,921
Corporation net income				6,119	8,559	9,328	12,566	11,965
Business privilege tax				8,407	11,405	10,244	9,234	8,956
Property	1,410	1,476	1,353	1,463	1,468	1,730	2,605	2,625
General	847	829	864	882	891	1,093	1,222	1,248
Severance	3,166	3,306	3,140	4,358	6,880	12,594	15,418	18,619
Oil and gas production	3,166	3,306	3,140	4,358	6,880	8,283	9,288	10,730
Coal production						4,311	6,130	7,889

Ohio

License			350,676	326,457	355,470	421,469	370,748	335,460
Public utilities							6,713	8,202
Coal consumption								341
Strip mining admin.			698	729	596	772	804	741
Individual income		111,269	373,543	419,174	481,785	511,636	614,879	775,494
Refunds				-51,681	-54,730	-64,683	-69,558	-77,048
Gross collections				470,855	536,515	576,319	684,437	852,542
Corporation net income		134,698	167,970	190,584	267,315	265,052	315,481	461,393
Property: special	60,553	65,515	75,950	84,219	91,335	98,161	108,347	119,617

Table A–1 continued

Tax	1971	1972	1973	1974	1975	1976	1977	1978 (prelim)
Ohio continued								
Severance		864	4,141	4,101	3,892	3,930	3,918	3,800
Coal and salt		555	2,136	2,012	1,928	2,070	1,997	1,937
Oil		43	250	233	234	256	266	258
Oklahoma								
Individual income (net)	63,648	97,759	105,054	120,773	176,208	200,998	216,833	252,127
Refunds					−24,837	−33,514	−56,263	−60,823
Gross collections					201,045	234,512	273,096	312,950
Corporation net income	25,207	28,014	35,434	40,366	46,053	53,430	70,635	91,375
Severance	51,280	73,342	71,456	96,980	128,096	151,316	191,351	230,368
Gross production	50,099	72,164	70,326	95,898	126,858	150,071	189,180	215,925
Gas conservation								11,807
Petroleum excise	1,181	1,178	1,130	1,082	1,238	1,245	2,171	2,636
Oregon								
Individual income (net)	226,245	251,226	300,555	352,396	427,002	472,147	561,895	686,248
Refunds				−64,595	−69,547	−77,540	−87,107	−95,405
Gross collections				416,991	496,549	549,687	649,001	781,653
Corporation net income	24,517	40,606	51,131	85,734	90,691	66,657	91,104	125,474
Severance	2,538	2,381	2,581	2,824	3,084	3,458	3,680	4,117
Eastern and western								
Oregon severance					1,880	2,141	1,917	2,157

Pennsylvania

	1	2	3	4	5	6	7	8
License						593,323	664,741	726,844
Service mining conservation						387	409	421
Individual income (net)	135,067	730,641	1,010,825	1,115,612	995,409	1,062,210	1,178,071	1,327,816
Refunds				−9,050	−38,098	−28,360	−24,837	−18,783
Gross collections				1,124,662	1,033,507	1,090,570	1,202,909	1,346,599
Corporation net income	431,696	481,600	497,212	540,103	601,016	616,872	665,993	786,976
Property: special	33,765	35,672	39,963	46,046	47,881	60,054	62,524	69,622
Utility property	30,840	32,307	36,317	42,426	43,731	55,290	57,527	64,967
Domestic corporation				3,415	3,893	4,495	4,250	4,401
Foreign corporations				235	257	269	747	253

South Dakota

	1	2	3	4	5	6	7	8
Severance						310	536	872
Mineral and mineral products								526
Oil and gas						310	536	346

Tennessee

	1	2	3	4	5	6	7	8
License							156,017	164,284
Strip mining permits							193	376
Individual income (dividends and interest tax) (net)	12,383	13,598	15,103	16,464	18,436	22,131	22,385	24,857
Refunds					−62	−49	−119	−110
6% tax (gross)					14,950	18,011	18,645	21,181
4% tax (gross)					3,503	3,880	3,713	3,699
Penalties and interest					45	289	147	87
Corporation net income	59,455	77,804	102,978	112,974	126,715	128,621	156,042	170,848
Excise (income)						128,621	150,935	164,877
Severance: coal tax				810	1,595	1,818	2,052	2,128

Table A–1 continued

Tax	1971	1972	1973	1974	1975	1976	1977	1978 (prelim)
Texas								
Property: general	63,837	61,589	57,191	50,811	44,901	36,668	42,755	44,598
Severance	307,924	311,979	339,757	523,745	666,876	800,693	907,281	959,686
Natural and casing								
head gas	108,809	114,380	124,902	171,068	257,325	364,588	474,318	517,844
Crude oil	192,474	190,785	207,522	344,832	402,553	429,105	426,373	435,223
Sulfur	4,291	4,611	4,959	5,516	4,787	4,790	4,480	4,636
Oil and gas regulation	2,350	2,203	2,374	2,329	2,211	2,210	2,110	1,983
Utah								
Individual income (net)	61,884	74,096	88,547	90,032	104,919	140,562	153,562	188,894
Refunds				−17,668	−16,372	−17,922	−25,406	−31,030
Gross collections				107,700	121,291	158,484	183,674	219,924
Corporation net income (franchise)	11,085	12,636	29,575	20,173	18,002	24,501	24,866	29,448
Property: General	13,089	14,634	9,492	3,409	258	204	197	186
Severance	4,671	3,938	3,913	5,292	6,238	11,723	8,931	8,926
Mine occupation	3,374	2,506	2,383	2,871	342	4,731	2,490	2,283
Oil and gas production	1,297	1,432	1,530	2,421	5,896	6,992	6,441	6,643
Virginia								
License						124,741	129,856	126,603
Strip mining permits						309	373	352

Individual income (net)	312,984	365,984	441,900	468,967	547,125	614,575	714,086	874,817
Refunds					NA	−137,149	−148,136	−171,553
Gross collections					NA	751,724	862,222	1,046,370
Corporate net income	64,705	77,642	96,618	106.406	117,065	130,417	159,152	164,790
Washington								
Property	116,525	132,350	131,785	91,245	156,641	271,528	303,165	349,229
General	66,957	76,064	72,571	27,715	87,530	188,254	205,795	234,988
Wisconsin								
Individual income (net)	507,146	594,697	727,885	802,995	873,723	959,923	1,144,073	1,324,679
Refunds				−98,461	−100,806	−111,377	−122,548	−156,220
Gross collections				901,456	974,530	1,071,300	1,266,621	1,480,898
Corporation net income	88,792	116,805	135,107	160,269	153,407	190,419	251,657	284,979
Property	86,386	93,054	91,295	92,087	126,707	91,910	100,874	
Severance: production tax (timber/iron ore)	261	313	475	421	504	651	698	602
Wyoming								
Property	9,725	9,660	7,570	9,797	6,315	7,045	9,049	17,589
General	8,537	8,490			6,115	6,659	8,794	17,355
Severance	4,877	5,075	5,307	5,086	18,543	40,974	46,969	66,021
Mineral excise, 4%					NA	38,791	43,732	44,116
Mineral excise, 2%					NA	1,525	1,925	3,766
Coal						283	997	17,716
Coal and gas conversion					307	375	315	423
Coal and gas production	4,877	5,075	5,307	5,086				

Bibliography

Agria, S.R. "Special Tax Treatment of Mineral Industries." In A.C. Harberger and M.J. Bailey, eds., *The Taxation of Income from Capital.* Washington: Brookings, 1969.

Agterberg, F.P., and A.M. Kelly. "Geomathematical Methods for use in Prospecting." *Canadian Mining Journal* (May 1971).

Allais, M. "Method of Appraising Economic Prospects of Mining Exploration over Large Territories: Algerian Sahara Case Study." *Management Science* 3 (July 1957): 285-347.

Atkinson, A.B., and J. Stiglitz. "The Design of Tax Structure: Direct versus Indirect Taxation." *Journal of Public Economics* 6 (1976): 55-76.

Beasley, Charles A., Charles R. Tatum, and Brian W. Laurence. "A Program for the Determination of the Technical and Economic Feasibility of Mining Operations." In T.B. Johnson and Donald W. Gentry, eds., *12th Annual International Symposium on the Applications of Computers in the Mineral Industry.* Golden, Colo.: Colorado School of Mines, 1974.

Bennett, Harold J. et al. *Financial Evaluation of Mineral Deposits Using Sensitivity and Probabilistic Methods.* Washington: U.S. Department of the Interior, Bureau of Mines, 1970.

Binger, Brian R. "Long-Run Costs in Uranium Production." University of Chicago, April 1977.

Boyle, Gerald J. "Taxation of Uranium and Steam Coal in the Western States." In A.M. Church, ed., *Non-Renewable Resource Taxation in the Western States.* Cambridge: Lincoln Institute, 1977.

Brown, G.A. "The Evaluation of Risk in Mining Ventures." Canadian Institute of Mining Trans. 72, 1970 (*CIM Bulletin,* October 1970).

Burness, H.S. "On Taxation of Nonreplenishable Natural Resources." *Journal of Environmental Economics and Management* 3 (1976): 289-311.

Burt, O.R., and R.G. Cummings. "Production and Investment in Natural Resource Industries." *American Economic Review* 60 (September 1970): 576-590.

Byrne, R.F., and L.J. Sparvero. "A Study and Model of the Exploration Process in the Non-Fuel Mineral Industry." Report to

U.S. Bureau of Mines, 1969. Available through Clearinghouse for Federal Scientific and Technical Information, Springfield, Virginia, PB 188 526.

Carlisle, D. "The Economics of a Fund Resource with Particular Reference to Mining." *American Economic Review* 44 (September 1954): 595–616.

Cigno, A., and B. Hool. "Bounded Rationality and Intertemporal Decision-Making." *Metroeconomica,* 1980.

Conrad, Robert F. "Taxation and the Theory of the Mine." Ph.D. thesis, University of Wisconsin, 1978*a*.

———. "Royalties, Cyclical Prices and the Theory of the Mine." *Resources and Energy* 1 (1978*b*): 139–150.

———. "Mining Taxation: A Numerical Introduction." Duke University, 1979*a*.

———. "Taxation of the Concentration Decision." Duke University, 1979*b*.

Conrad, Robert F., and Bryce Hool. "A Theory of the Mine." Duke University, 1979*a*.

———. "Resource Taxation with Heterogeneous Quality and Endogenous Reserves." Duke University, 1979*b*.

Convery, F., and R.F. Conrad. *Irish Minerals Policy.* Government of Ireland, 1979.

Cooper, D.O., L.B. Davidson, and K.M. Reim. "Simplified Financial and Risk Analysis of Minerals Exploration." In John R. Sturguled, ed., *11th International Symposium on Computer Applications in the Mineral Industry.* Tucson: University of Arizona, 1973.

Davidson, P. "The Depletion Allowance Revisited." *Natural Resources Journal* 1 (1970).

Dougherty, E.L., and R.W. Schewel. "Computer Exploration Techniques." *Mining Magazine* 123 (August 1970).

Fiekowsky, S., and A. Kaufman. "Mineral Taxation." in W.A. Vogeley and H.E. Risser, eds., *Economics of the Mineral Industries,* 3d ed. New York: AIME, 1976.

Frohling, Edward S. "What Kind of Financing You Need to Get Your Small Mine into Production." *Engineering and Mining Journal* (December 1970).

Frohling, Edward S., and Robert M. McGeorge. "How Stepwise Financing Can Turn Your Prospect into an Operation Mine." *Mining Engineering* 27 (September 1975): 30–32.

Bibliography

Gaffney, M., ed. *Extractive Resources and Taxation.* Madison: University of Wisconsin, 1967.

Gentry, D.W. "Two Decision Tools for Mining Investment and How to Make the Most of Them." *Mining Engineering* 23 (November 1971): 55–58.

Gentry, D.W., and T.S. O'Neill. "A Short Course on Financial Modeling and Evaluation of New Mine Properties." Golden, Colo.: 12th APCOM Symposium, 1974.

Gillis, S. Malcolm. "Severance Taxes on North American Energy Resources: A Tale of Two Minerals." *Growth and Change* 10 (January 1979): 55–71.

Gillis, S.M., and C.E. McClure, Jr. "The Incidence of World Taxes on Natural Resources, with Special Reference to Bauxite." *American Economic Review* 65 (May 1975).

Gillis, M. et al. *Taxation and Mining.* Cambridge: Ballinger, 1978.

Griffin, K., and R.B. Shelton. "Coal Severance Tax Policies in the Rocky Mountain States." *Policy Studies Journal* 7 (1978): 29–40.

Halls, J.L., P.P. Bellum, and C.K. Lewis. "The Determination of Optimum Ore Reserves and Plant Size by Incremental Financial Analysis." *AIME—Society of Mining Engineers.* Preprint, February 1969.

Hansen, C.J. "If Cabbages Were Kings: A Practical Approach to the Taxation of Mining Properties." In A. Church, ed., *Non-Renewable Resource Taxation in the Western States.* Cambridge: Lincoln Institute, 1977.

Harberger, A.C. "The Taxation of Mineral Industries." In A.C. Harberger, ed., *Taxation and Welfare.* Boston: Little, Brown and Co., 1974.

Harris, DeVerle D. "Risk Analysis in Mineral Investment Decisions." *AIME—Society of Mining Engineers Transactions.* 247 (September 1970): 193–201.

Hazen, S.W. "Some Statistical Techniques for Analysis of Mine and Mineral Deposit Sample and Assay Data." *Bureau of Mines Bulletin* no. 621, 1967.

Hoskold, H.D. *The Engineers' Valuing Assistant.* London: Longmans, Grum and Co., 1877.

Hotelling, H. "The Economics of Exhaustible Resources." *Journal of Political Economy* 39 (April 1931): 137–175.

Just, Evan. "The Production Process." In W.A. Vogely and

H.E. Risser, eds., *Economics of the Mineral Industries,* 3d ed. New York: AIME, 1976.

Koch, G.S. "Statistical Analysis of Assay Data from the Round Mountain Silver Prospect, Custer County, Colorado." Washington: Bureau of Mines Report of Investigations, R.I. 7486, 1971.

Koch, G.S., and R.F. Link. "Sampling Gold Ore by Diamond Drilling in the Homestead Mine, Load, S. Dakota." U.S. Bureau of Mines, R.I. 7508, 1971.

Lane, K.F. "Choosing the Optimum Cut-off Grade." *Quarterly of the Colorado School of Mines* 59 (October 1964): 811–830.

Lewis, F. Milton, and Roshan B. Bhappu. "Evaluating Mining Ventures via Feasibility Studies." *Mining Engineering* 27 (October 1975): 48–54.

Lindley, A.H. et al. "Mineral Financing." In W.A. Vogely and H.E. Risser, eds., *Economics of the Mineral Industries,* 3d ed. New York: AIME, 1976.

Link, A.A. "Political Constraints and North Dakota's Coal Severance Tax." *National Tax Journal* 31 (September 1978): 263–268.

Link, R.F. et al. "Statistical Analysis of Gold Assay and Other Trace Element Data." U.S. Bureau of Mines, R.I. 7495, 1971.

Lockner, A.O. "The Economic Effect of a Progressive Net Profit Tax on Decision-Making by the Mining Firm." *Land Economics* 38 (November 1962): 341–349.

———. "The Economic Effect of the Severance Tax on the Decisions of the Mining Firms." *Natural Resources Journal* 4 (January 1965): 468–485.

Long, Stephen C.M. "Coal Taxation in the Western States: The Need for a Regional Tax Policy." *Natural Resources Journal* 16 (April 1977).

McClure, Charles E., Jr. "State Corporate Income Tax: Lambs in Wolves' Clothing?" OTA Paper 25, Office of Tax Analysis, U.S. Treasury, March 1977.

———. "Interstate Exporting of State and Local Taxes: Estimates for 1962." *National Tax Journal* 20 (March 1967): 49–77.

———. "The Interregional Incidence of General Regional Taxes." *Public Finance* 24 (1969): 457–483.

———. "Economic Constraints on State and Local Taxation of Energy Resources." *National Tax Journal* 31 (1978): 257–262.

McDonald, Stephen L. "The Non-Neutrality of Corporate Income

Taxation: A Reply to Steiner." *National Tax Journal* 17 (March 1964): 101–104.

———. "Incentive Policy and Supplies of Energy Sources." *American Journal of Agricultural Economics* 56 (May 1974): 402.

———. "The Effects of Severance vs. Property Taxes on Petroleum Conservation." *Proceedings of the National Tax Association* (1965): 320–327.

———. "Percentage Depletion, Expensing of Intangibles, and Petroleum Conservation." In M. Gaffney, ed., *Extractive Resources and Taxation*. Madison, University of Wisconsin, 1967.

———. "Taxation System and Market Distortion." In William A. Vogely and Robert J. Kalter, eds., *Energy Supply and Government Policy*. Ithaca: Cornell University Press, 1976.

———. *Federal Tax Treatment of Oil and Gas*. Washington: Brookings, 1963.

———. "Percentage Depletion and Tax Neutrality: A Reply to Messrs. Musgrave and Eldridge." *National Tax Journal* 15 (September 1962):314–326.

———. "Percentage Depletion and the Allocation of Resources: The Case of Oil and Gas." *National Tax Journal* 14 (December 1961): 323–336.

McGeorge, Robert L. "Approaches to State Taxation of the Mining Industry." *Natural Resources Journal* 10 (January 1970): 156–170.

MacKenzie, B.W. et al. "The Effect of Uncertainty on the Optimization of Mine Development." In T.B. Johnson and Donald W. Gentry, eds., *12th Annual International Symposium on the Applications of Computers and Mathematics in the Minerals Industry*. Golden, Colo.: Colorado School of Mines, 1974.

McKinstry, H.E. *Mining Geology*. New York: Prentice-Hall, 1948.

Maximov, A. et al. *Short Course in Geological Prospecting and Exploration*. Moscow: MIR Publishers, 1973 (1st English translation).

Maxwell, J.A. *Financing State and Local Governments*. Washington: Brookings, 1965.

Mieszkowski, A. "The Property Tax: An Excise Tax or a Profits Tax?" *Journal of Public Economics* 1 (1972): 73–96.

Patterson, J.A. "Estimating Ore Reserves Following Logical Steps." *Engineering and Mining Journal* 160 (September 1959).

Peterson, F.M. "The Long Run Dynamics of Minerals Taxation." University of Maryland, 1976.

Peterson, F.M., and A.C. Fisher. "The Exploitation of Extractive Resources: A Survey." *Economic Journal* 87 (December 1977): 681–721.

Pfluder, E.T., ed. *Surface Mining.* New York: AIME, 1968.

Preston, Lee E. *Exploration for Non-Ferrous Metals.* Baltimore: Johns Hopkins, 1960.

Roff, Arthur W., and James C. Franklin. "A Statistical Mine Model for Cost Analysis, Planning and Decision Making." *Quarterly of the Colorado School of Mines* 59 (October 1964): 915–924.

Shelton, R.B., and W.E. Morgan. "Resource Taxation, Tax Exportation and Regional Energy Policies." *Natural Resources Journal* 17 (April 1977): 261–282.

Solow, R. "The Economics of Resources and the Resources of Economics." *American Economic Review* (May 1974).

Sorenson, J.B., and R. Greenfield. "New Mexican Nationalism and the Evolution of Energy Policy in New Mexico." *Natural Resources Journal* 17 (April 1977): 287–292.

Steering Committee on the Impact of Taxation on Energy Markets. "A Taxonomy of Energy Taxes." Washington: National Academy of Sciences, 1979.

Steiner, P.O. "Percentage Depletion and Resource Allocation." In *Tax Revision Compendium.* Washington: GPO, 1959.

Stinson, Thomas F. "State Taxation of Mineral Deposits and Production." Office of Research and Development, U.S. Environmental Protection Agency. Washington: GPO, 1977.

————. "State Taxation of Mineral Deposits and Production." Rural Development Report no. 2. Washington: USDA, 1978.

Thomas, L.J. *An Introduction to Mining.* Sydney: Hicks, Smith and Sons, 1973.

Thomas, E.G. "Justification of the Concept of High-Grading Metalliferrous Ore Bodies." *Mining Magazine* 134 (May 1976): 393–396.

Truscott, S.J. *Mine Economics,* 3d ed., revised by J. Russell. London: Mining Publications, 1962.

Vogely, W.A., and H.E. Risser, eds. *Economics of the Mineral Industries,* 3d ed. New York: AIME, 1976.

Walduck, G.P. "Justification of the Concept of High-Grading in the

Metalliferous Ore Bodies: A Dissenting View." *Mining Magazine* 134 (1976): 65-66.

Warren, R. *State Mineral Taxation*. Cambridge: Harvard, 1944.

Yasnowsky, P.N., and A.P Graham. "State Severance Taxes on Mineral Production." *Proceedings of the Council of Economics, 105th Annual Meeting*. American Institute of Mining, Metallurgical, and Petroleum Engineers, Inc., Las Vegas, 22-26 February 1976.

Index

Index

About the Authors

Robert F. Conrad is currently assistant professor of economics at Duke University. He received the Ph.D. in economics at the University of Wisconsin, Madison in 1978. Prior to his current position, Professor Conrad was a visiting lecturer in economics at Northwestern University and worked in the office of international tax affairs at the U.S. Treasury Department.

Professor Conrad has served as a consultant on mineral taxation policy for the governments of Canada, Ireland, and Indonesia. His other research interests include capital taxation, taxation of multinational firms, and health economics.

Bryce Hool received the Ph.D. in economics from the University of California at Berkeley, after receiving the B.Sc. in mathematics and M. Comm. in economics from the University of Canterbury. He is currently an associate professor of economics at the State University of New York at Stony Brook and was an assistant professor at the University of Wisconsin at Madison. His other published research is in the areas of general equilibrium theory, monetary theory, and macroeconomic theory and policy.